Mechanisms and Machine Science

Volume 128

This book series establishes a well-defined forum for monographs, edited Books, and proceedings on mechanical engineering with particular emphasis on MMS (Mechanism and Machine Science). The final goal is the publication of research that shows the development of mechanical engineering and particularly MMS in all technical aspects, even in very recent assessments. Published works share an approach by which technical details and formulation are discussed, and discuss modern formalisms with the aim to circulate research and technical achievements for use in professional, research, academic, and teaching activities.

This technical approach is an essential characteristic of the series. By discussing technical details and formulations in terms of modern formalisms, the possibility is created not only to show technical developments but also to explain achievements for technical teaching and research activity today and for the future.

The book series is intended to collect technical views on developments of the broad field of MMS in a unique frame that can be seen in its totality as an Encyclopaedia of MMS but with the additional purpose of archiving and teaching MMS achievements. Therefore, the book series will be of use not only for researchers and teachers in Mechanical Engineering but also for professionals and students for their formation and future work.

The series is promoted under the auspices of International Federation for the Promotion of Mechanism and Machine Science (IFToMM).

Prospective authors and editors can contact Mr. Pierpaolo Riva (publishing editor, Springer) at: pierpaolo.riva@springer.com

Indexed by SCOPUS and Google Scholar.

Juan Carlos García Prada · Cristina Castejon ·
Jose Ignacio Pedrero Moya
Editors

Trends in Educational Activity in the Field of Mechanism and Machine Theory (2018–2022)

Selected Papers from ISEMMS 2022

IFToMM

Springer

Editors
Juan Carlos García Prada
Department of Mechanical Engineering
Universidad Nacional de Educatión
a Distancia (UNED)
Madrid, Spain

Jose Ignacio Pedrero Moya
Department of Mechanical Engineering
Universidad Nacional de Educatión
a Distancia (UNED)
Madrid, Spain

Cristina Castejon
Department of Mechanical Engineering
Carlos III University of Madrid
Madrid, Spain

ISSN 2211-0984 ISSN 2211-0992 (electronic)
Mechanisms and Machine Science
ISBN 978-3-031-25729-2 ISBN 978-3-031-25730-8 (eBook)
https://doi.org/10.1007/978-3-031-25730-8

This Springer imprint is published by the registered company Springer Nature Switzerland AG
The registered company address is: Gewerbestrasse 11, 6330 Cham, Switzerland

Preface

The International Symposium on the Education in Mechanism and Machine Science (ISEMMS 2021-22) is the third event of the series of symposia focused on mechanism and machine science (MMS) education, as continuation of the first seminar about "Education in MMS" proposed in the seventies of last century by Prof. Artobolevsky.

The conventional engineering teaching is currently enhanced with the new technologies and methods included by and for the new generations. The goal is to obtain graduates with relevant competences from industry stakeholders or research institutions. Today, with the COVID pandemic, there is a need to share and discuss the current problems in MMS education with the aim of providing solutions and identifying appropriate trends for a modern world common vision in the Engineering education field.

This third ISEMMS tried to establish a forum where teachers and researchers interchange experiences in the field of MMS. After the review process, 24 papers were selected for presentation in this symposium. The program of ISEMMS 2021-22 had six technical sessions with oral and online presentations that allowed sharing experiences and discussions. The contributions cover topics related with MMS in the engineering program: methodology, new trends in Mechanical Engineering education, experiences in Mechanical Engineering education, MMS engineering program: virtual labs, and MMS experiences in Pandemic times. We thank the authors for their excellent and interesting contributions with new ideas, procedures, and methodologies for the MMS educator's community.

During the conference, we were honored with the talk of the Dean of the School of Engineering of the UNED, Prof. Cristina González Gaya, in the plenary session "Approach to the Education in MMS at the UNED School of Engineering." The speech offered to the assistants the routes toward the university education in future.

The editors would like to express our grateful thanks to the members of the scientific committee for ISEMMS 2021-22 who have collaborated for the success of the symposium, especially professors: M. Ceccarelli (Italy), G. Carbone (Italy), P. Fanghella (Italy), P. Flores (Portugal), A. Giménez (Spain), E. Krylov (Russia), L. Martínez-Muneta (Spain), J. Mayo (Spain), Karol Miller (Australia), L. Jordi (Spain), V. Petuya (Spain), R. Rubini (Italy), and M. Tomás-Rodríguez (UK).

We would also like to express grateful thanks to the members of the local organizing committee of ISEMMS 221-22 who have to work very hard to make possible the event, especially Prof. A. Bustos.

Thanks for the financial support to UNED (National University of Distance Education), IFToMM, AEIM (Spanish Association of Mechanical Engineering), and the Department of Mechanics of UNED.

Finally, we would like to remark on the invaluable support of President of IFToMM, Prof. Andrés Kecskemethy, during all the stages of the symposium and the continuing help to achieve with success the ISEMMS series.

Madrid, Spain — Juan Carlos García Prada
Madrid, Spain — Cristina Castejon
Madrid, Spain — Jose Ignacio Pedrero Moya

Contents

Part I
Mechanism and Machine Science in the Engineering Program: Methodology

Chapter 1
Teaching Concept for the Practical Application of Product Development and Project Management Methods

Thomas Knobloch, Nils Mandischer, Judith Merz, Jan Wiartalla, Mathias Hüsing, Burkhard Corves, Sebastian Lüke, Stefan Kurtenbach, and Jascha Paris

Abstract In the renewed master's program in product development at RWTH Aachen University, a new teaching concept has been introduced with the course "Applied Design and Product Development I and II". Students can apply the knowledge they have acquired during their studies in the fields of design, product development and project management in an industry-oriented manner. Together with an industry partner from the mechanical engineering industry who provides a design assignment, the product development process takes place over the course of a year—from the blank sheet of paper to a finished prototype. The industrial partner benefits

T. Knobloch (✉) · N. Mandischer · J. Merz · J. Wiartalla · M. Hüsing · B. Corves
Institute of Mechanism Theory, Machine Dynamics and Robotics, RWTH Aachen University, Eilfschornsteinstr. 18, 52062 Aachen, Germany
e-mail: knobloch@igmr.rwth-aachen.de

N. Mandischer
e-mail: mandischer@igmr.rwth-aachen.de

J. Merz
e-mail: merz@igmr.rwth-aachen.de

J. Wiartalla
e-mail: wiartalla@igmr.rwth-aachen.de

M. Hüsing
e-mail: huesing@igmr.rwth-aachen.de

B. Corves
e-mail: corves@igmr.rwth-aachen.de

S. Lüke · S. Kurtenbach
TRAPO GmbH, Industriestr. 1, 48712 Gescher-Hochmoor, Germany
e-mail: slueke@trapo.de

S. Kurtenbach
e-mail: skurtenbach@trapo.de

J. Paris
Demcon Systec Industrial Systems GmbH, Wilhelm-Schickard-Str. 6, 48149 Münster, Germany
e-mail: research@jaschaparis.de

J. C. García Prada et al. (eds.), *Trends in Educational Activity in the Field of Mechanism and Machine Theory (2018–2022)*, Mechanisms and Machine Science 128,
https://doi.org/10.1007/978-3-031-25730-8_1

from the students' knowledge and performance thanks to their advanced academic background and their ability to view the project from a different perspective. This results in a practical experience for the students. Thus, the course "Applied Design and Product Development I and II" represents a synergy for the university students and industrial researchers. In this paper, the teaching concept is presented based on the project task from 2020.

Keywords Teaching · Design · Product development

1.1 Concept

Since 2020, the master's program in Product Development at RWTH Aachen University provides students with in-depth methodological knowledge of model-based product development. The course "Applied Design and Product Development I and II" makes a significant contribution to this by enabling students to put the knowledge they have acquired during their academic career into practice. The knowledge gained in the areas of design, product development, communication and project management can be applied in this course [1–4]. A brief overview of the content is shown in Fig. 1.1.

The main goal is to solve a task from a blank sheet of paper to a functional prototype set by an industrial partner. During the product development process, students are supervised and assisted by the industrial partner and experts from the responsible institute. The students inform the industrial partner in the form of short presentations about the current state of development. In this way, the students prepare themselves for the practical work of an engineer in product development. The opportunity for students to look at the project from a different perspective leads to beneficial results for the industrial partner.

TRAPO, based in Gescher-Hochmoor, Germany, has been a manufacturer of machines and complete systems for intralogistics for over 60 years. The company

Fig. 1.1 Contents of the course

offers system solutions in the field of conveying technology, packaging, palletizing, driverless transport systems and truck loading systems. The following illustrates the process based on a project task set by TRAPO in 2020.

1.2 Structure of the Course Based on an Example

The Applied Design and Product Development I and II course is scheduled for two semesters and begins in the summer semester. During each semester, three of the six milestones are covered according to the product development process as per VDI 2221 [5], as can be seen in Fig. 1.2.

A very important element of the evaluation of the students' results are six so-called Design Reviews, which follow the six milestones. There is feedback from the professors and the industry partner per student but also for the entire work. The Design Reviews are also examination components.

The course begins with the teaching and processing of the theoretical foundations of design methodology as well as project management required for the later course. Students can then choose one of the projects supervised by industrial partners together with one of six to eight institutes of the Faculty of Mechanical Engineering at RWTH Aachen University. The project tasks are provided directly from the industrial partners. Up to 15 students are involved in one project. At the beginning of the project phase, participants receive instructions from experts from the institute in charge of the project and from the industrial partner. The students will be divided into smaller groups in the next stage. In this way, either the same problem can be solved in several groups, or the subtasks of a higher-level question can be solved by assigning varying priorities to each group, which are defined by the industry partner. During the project processing, support is mainly provided by experts from the responsible

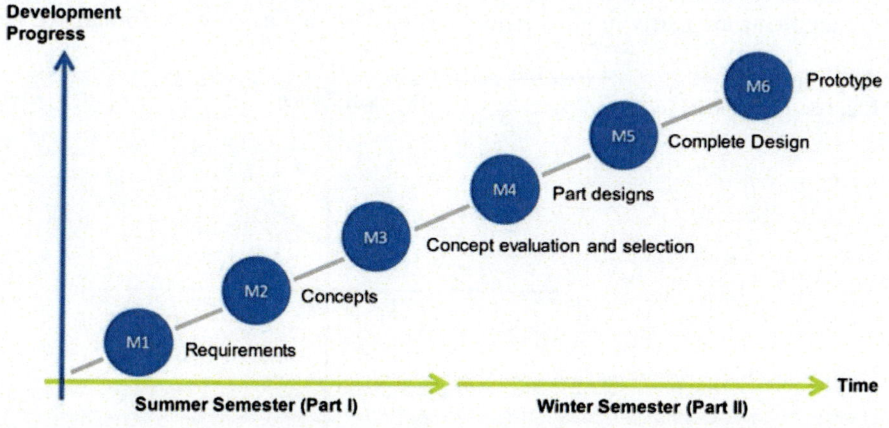

Fig. 1.2 Milestones of the course

Table 1.1 Conveying system requirements from TRAPO

Subject	Requirement
Conveyed materials	Packages, trays, boxes, cartons
Dimensions	Approx. 100 mm × 100 mm up to 800 mm × 400 mm
Conveyed weight	10–15 kg
Cycle time	1 s
Speed	2–3 m/s

institute. The design reviews are not only a key examination component but are very suitable means of informing the industry partner about the results.

The project task in 2020 was to look at the conveying technology of future intralogistics. Current conveying systems take up a lot of space and are located on different levels. Automated guided vehicles eliminate these problems because they are no longer a fixed conveying system, areas can be used multiple times and goods can be distributed flexibly. However, driverless transport systems have significantly lower performance, extensive safety technology is required, and the components are relatively expensive. Therefore, new conveying systems are to be developed. For the students to provide results as diverse as possible, three groups of five students each were formed. Each group was given specific instructions by TRAPO. The first group was to develop a conveying system with minimal infrastructure. The second group was to develop a conveying system that has a high degree of reconfigurability. The third group was to work on the development of a guideless conveying technology. The requirements that students receive from TRAPO are listed in Table 1.1.

However, these requirements are not the only ones imposed on the product. Students learn that in addition to the requirements set by the customer, other requirements for the product must be determined. According to Kano there are different types of requirements, which are shown through Fig. 1.3 [6].

The product features are categorized as Basic, Performance and Exciters. Basic requirements are partly implicit requirements. They are usually not formulated but

Fig. 1.3 Kano model [6]

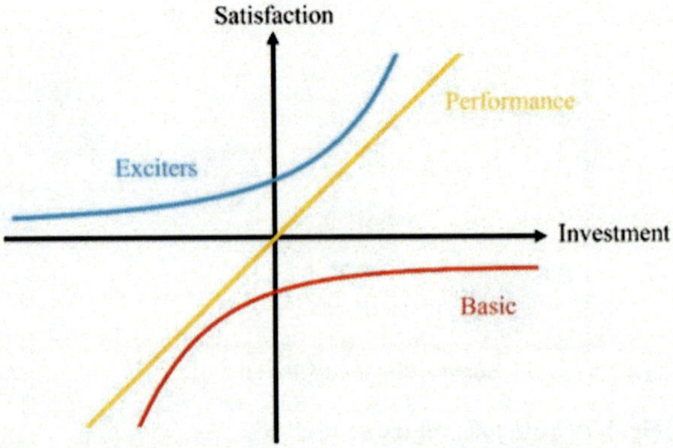

taken for granted. Basic requirements are most important for customer satisfaction and are critical for success. Performance requirements, on the other hand, are explicitly desired by the customer and are used by the customer for comparison with competitors. Exciting requirements are not clearly articulated and are also often not recognized by the customer. They are a key product differentiator. Students typically receive performance requirements as they received from TRAPO. However, the basic requirements and attractiveness requirements are not specified and must be determined independently. The first work package (M1) is also characterized by an extensive literature and patent research. The participants study the literature and learn about the state of the art. Another subtask of the first work package is the formulation of an offer letter to the industry partner. Students must create a fictitious company that the industry partner will hire to carry out the project. The offer should consist of the project objectives, approach, project time frame, and cost and resources required.

In the concept phase (M2), conceptual solutions are created based on the requirements. First, a functional structure of the product must be created. The individual subfunctions can then be assigned to their function carriers, i.e., components or assemblies of the product that perform this function [2]. As an example, Fig. 1.4 shows part of the functional structure of Group 1, which refers to the minimal infrastructure of the new conveying system.

The functions of the lowest level are called elementary functions because they cannot be divided further.

In the next step, principle solutions from these elementary functions should be developed. Students can use different methods to find principle solutions, for example brainstorming, the theory of inventive problem solving (TRIZ) [7] or the use of catalogues such as the Koller catalogue [8]. Finally, the principle solutions must be assembled into a morphological box after a compatibility check was carried out [9]. The technically feasible principle solutions should be combined in the morphological box to form complete solutions as shown in Fig. 1.5. The combination of partial solutions results in total solutions.

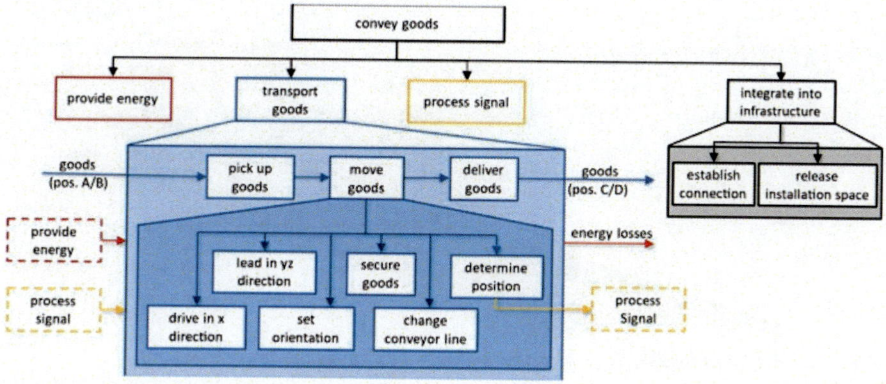

Fig. 1.4 Function structure "transport goods"

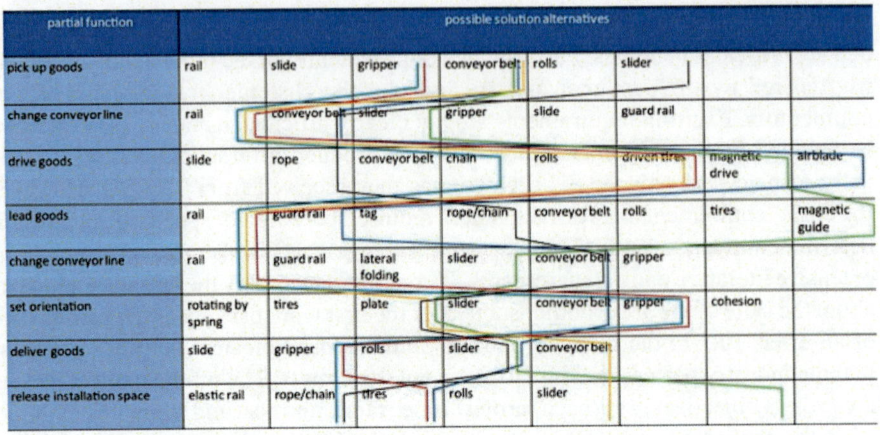

partial function	possible solution alternatives						
pick up goods	rail	slide	gripper	conveyor belt	rolls	slider		
change conveyorline	rail	conveyor belt	slider	gripper	slide	guard rail		
drive goods	slide	rope	conveyor belt	chain	rolls	driven tires	magnetic drive	airblade
lead goods	rail	guard rail	tag	rope/chain	conveyor belt	rolls	tires	magnetic guide
change conveyorline	rail	guard rail	lateral folding	slider	conveyor belt	gripper		
set orientation	rotating by spring	tires	plate	slider	conveyor belt	gripper	cohesion	
deliver goods	slide	gripper	rolls	slider	conveyor belt			
release installation space	elastic rail	rope/chain	tires	rolls	slider			

Fig. 1.5 Morphological box

The third design review of the project (M3) includes an evaluation of the various overall solutions. Students are introduced to several evaluation methods, including techno-economic assessment and utility analysis. Based on these, participants must establish evaluation criteria and deal with uncertainties. The outcome of the evaluation will be one or two complete solutions that will be further developed in the next semester. Figure 1.6 shows the best rated concept from part I of the course in 2020.

It is an externally driven system consisting of vehicles that are variable in height and width. This allows vehicles in the low position without goods to drive under vehicles in the high position so that only one track is required for driving in both directions.

In the fourth stage (M4), students are introduced to agile working using GitLab [10]. Git is a versioning service for text file formats and will now be used for issue tracking. All 15 students are divided into two different project groups, Mechanical

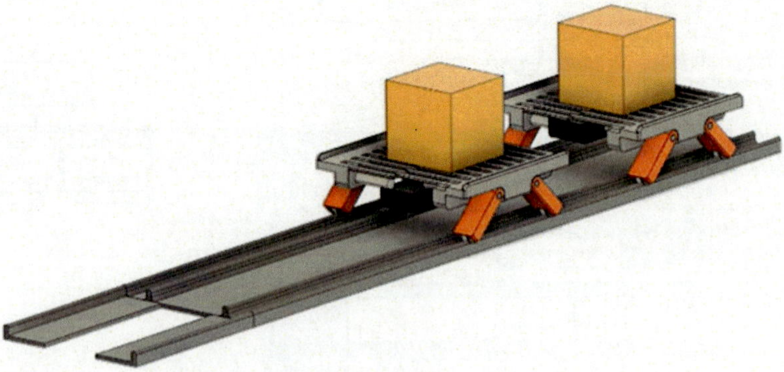

Fig. 1.6 Best rated concept at the end of part I

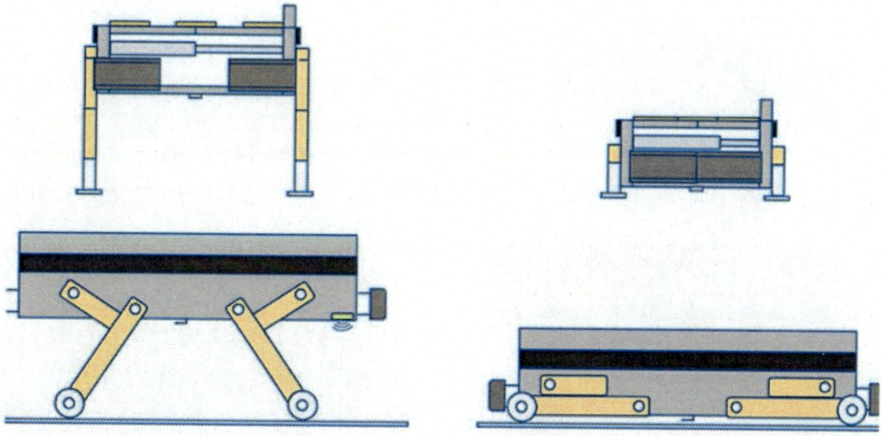

Fig. 1.7 Part design of the vehicle

Design and Mechatronics and Electronics. One person from each of the two project groups is also part of the project management group. The project management group must coordinate the other groups and is in direct contact with the industrial partner. They are also responsible for the management of the SCRUM and the creation of the relevant documents. The working group focusing on mechanical design is responsible for the design of mechanical parts and components, for deciding on the global design space as well as for design, load calculation and material selection. The Mechatronics and Electronics working group selects the mechatronic components and decides on the space to install the mechatronics. In addition, performance and demand calculations must be performed and the mechatronic components and their power supplies designed. The partial design of the vehicle developed by the Mechanical Design group is shown in Fig. 1.7.

The draft of the overall concept is to be completed by M5. The overall concept is defined by creating a final product structure for the overall system. All modules and components should be designed, and all interfaces developed. Furthermore, the manufacturing strategies of all components must be defined, which includes the decision for external or internal production and the manufacturing process. The status of the overall concept should also be analyzed against the requirements of design review 4. Next, a safety concept should be developed and a risk analysis in the form of an FMEA [11] or fault tree analysis [12] should be performed. Finally, the entire system will be fully mapped in CAD and the major components will be designed in detail. Figure 1.8 shows the complete design of the vehicle.

For the sixth design review (M6), the overall design and solution path should be presented. In addition, production documents of selected components are to be created, as well as a parts list and building structure. A cost breakdown and a cost analysis of the product should also be performed with a review of costs incurred, ABC analysis to determine cost drivers, and identification of cost reduction opportunities.

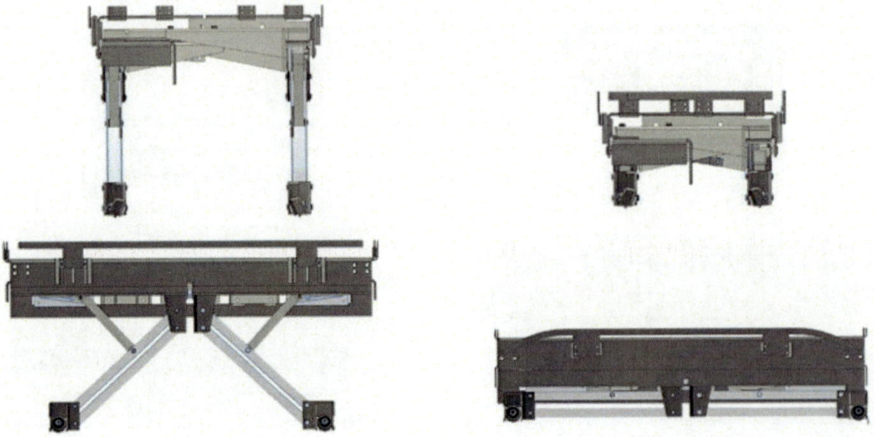

Fig. 1.8 Complete design of the vehicle

The prototype should be designed virtually or through 3D printing. Finally, a presentation of the product is made with animations and visualizations. The animation of the complete system is shown in Fig. 1.9.

At the end of each semester, students also prepare a final project report summarizing the results of the semester. Students receive their individual overall grade through a combination of the individual grades of the final project report and individual design reviews. At the end of the semester, the students gain insight into the company's operations and manufacturing processes by visiting the industrial partner and gain the opportunity to intensify their contacts to get to know about master thesis proposals and potential job offerings.

Fig. 1.9 Animation of the entire system

1.3 Summary

With the course "Applied Design and Product Development I and II" a modern teaching concept was created, which teaches students practical experience and allows them to implement a practice-related project together with an industrial partner. The industrial partner benefits from the students' knowledge and results, as they have an advanced academic background and can see the project from a different perspective. This gives students hands-on experience. In this way, the course "Applied Design and Product Development I and II" is a synergy between university students and researchers from industry.

References

1. Ehrlenspiel, K.: Integrierte Produktentwicklung, Denkabläufe, Methodeneinsatz, Zusammenarbeit, 4. Auflage. Hanser (2009)
2. Feldhusen, J., Gebhardt, B.: Product Lifecycle Management für die Praxis, Ein Leitfaden zur modularen Einführung. Springer, Umsetzung und Anwendung (2008)
3. Specht, G., Beckmann, C., Amelingmeyer, J.: F- & E-Management, Kompetenz im Innovationsprozess, 2. Auflage. Schäffer-Pöschel (2002)
4. Norm VDI 2221 Part 1: Design of technical products and systems – Model of product design, Verein Deutscher Ingenieure, Beuth-Verlag, Berlin (2019)
5. Pahl, G., Beitz, W., Feldhusen, J., Grote, K.H.: Engineering Design: A Systematic Approach, 3rd edn. Springer (2014)
6. Dalton, J.: Kano model. In: Great Big Agile. Apress, Berkeley, CA (2019). https://doi.org/10.1007/978-1-4842-4206-3_37
7. Petrov, V.: TRIZ. Theory of Inventive Problem Solving, Level 1. Springer Cham (2019)
8. Koller, R., Kastrup, N.: Prinziplösungen zur Konstruktion technischer Produkte, 2. Springer (1998)
9. Zwicky, F., Wilson, A.G.: New Methods of Thought and Procedure—Contribution to the Symposium of Methodologies. New York Inc., Springer (1967)
10. Choudhury, P., Crowston, K., Dahlander, L., et al.: GitLab: work where you want, when you want. J Org Design **9**, 23 (2020). https://doi.org/10.1186/s41469-020-00087-8
11. Ben-Daya, M.: Failure mode and effect analysis. In: Ben-Daya, M., Duffuaa, S., Raouf, A., Knezevic, J., Ait-Kadi, D.: Handbook of Maintenance Management and Engineering. Springer, London (2009)
12. Häring, I.: Fault tree analysis. In: Technical Safety, Reliability and Resilience. Springer, Singapore (2021)

Chapter 2
Implementation of Projects Oriented Learning in Mechanical Engineering Universities and Vocational Colleges

Enrique Soriano-Heras, Higinio Rubio, Jesús Manuel García-Alonso, and Abraham Vadillo

Abstract University Carlos III of Madrid (UC3M), Technical University of Madrid (UPM) and the vocational college Salesians Atocha (CSA), have been steadily working on teaching their subjects through the Project-Oriented Learning (POL) methodology. Later they also implemented more didactics methodologies to achieve higher development in the competences of students. These schools incorporated Service-Learning (SL) and Knowledge-Transfer (KT), by means of several tools such as Dynamic Technical Documentation (DTD), Google Suites and Moodle. Recently, other teaching methodologies have been added during COVID 19 pandemic, which pay special attention to units which are particularly difficult to understand, didactic methodologies such as "Kounaikenshuu", "Flipped Classroom", and tools such as "Portfolio", "Plickers" and "Kahoot". The implementation of these new didactic methodologies make possible not only improve students' academic training but also achieved the competencies marked for the Bachelor Degree in Mechanical Engineering at UC3M and UPM and for Mechanical Fabrication in CSA These results were also confirmed by the satisfaction surveys carried out by the collaborating partners of the projects which showed improvements of between 19 and 43% in theoretical training and between 53 and 68% in practical training at UPM, whereas in CSA, the improvement in theoretical training was 22 and 85% in practical training according to the criterion of the partner-collaborators.

Keywords Flipped classroom · Project oriented learning · Service learning and collaborative work

E. Soriano-Heras (✉) · H. Rubio · A. Vadillo
Department of Mechanical Engineering, University Carlos III of Madrid, Leganes, Spain
e-mail: esoriano@ing.uc3m.es

J. M. García-Alonso
ETSIDI, Technical University of Madrid, Madrid, Spain

J. C. García Prada et al. (eds.), *Trends in Educational Activity in the Field of Mechanism and Machine Theory (2018–2022)*, Mechanisms and Machine Science 128,
https://doi.org/10.1007/978-3-031-25730-8_2

2.1 Introduction

Today there are rapid changes and short-lived, in the various fields of innovation need therefore a model of teaching and learning versatile and flexible, quickly adapting to the needs the industrial world, going from Taylorism to a "social production" model where teamwork prevails, circles production, innovation, entrepreneurship, ...

The need to adapt to the new times and places of change is essential that universities change their way of proceeding (theory, practice, planning, implementation, thinking, acting) on another, more global, and comprehensive, in which enhances interdisciplinary, social and cultural skills, through a teaching–learning method according to the teachings transmitted [1–3].

Among the different existing active methodologies, Project Oriented Learning (POL), variant of Project Based Learning (PBL), is a methodology in which the faculty raises a project to a group of students getting them become more "active", responsible and committed to their learning [4–7]. To develop this learning methodology is needed from students, good management in various disciplines and sources of information, they will be needed when developing projects. In addition, it will be necessary, if not essential, coordination and involvement of various materials, enhancing the interdisciplinary nature of this learning methodology offering, to the students, an overview of the participating and transverse subjects and the teaching staff the possibility of working together, and with other departments, being in continuous learning and continuous improvement [8–10]. Flipped Classroom (FC), is a pedagogical model that transfers the work done so far in the classroom, outside, using classroom time, to facilitate and enhance skills [11].

The "Kounaikenshuu" is a Japanese methodology for collaborative professional development of consistent professors in a constant improvement of the teaching units, by the other fellow professors and other schools, through "Yugyou Kenkyuu" (lessons study) and "Yugyou Bunseki" (analysis lessons), such as, through continuous improvement and research of these teaching units [6, 12].

A mixture of POL, FC and "Kounaikenshuu" has been used in the work done at universities UC3M, UPM and at the vocational college CSA since the course 2014–15, for the subjects of Technical Office and Fabrication, respectively. This mix of methodologies based on constructivism supports the idea that people build learning through experience. Applying the hybrid methodology (POL-FC-"Kounaikenshuu"), students from the beginning must work at home, preparing lessons, both individually and as a team, consult bibliography, both in the library and through computers, be able to contrast the information found in order to plan and develop the action to be performed, to further evaluate the achievements. All these steps will lead students moving in a continuously activity. Once the project is completed, the team, made a self-critical reflection on knowledge achieved in each of the participating subjects [8].

Teachers who choose to use this methodology should consider several measures to ensure that their application is successful:

- The objectives to be achieved with the project.
- The technical knowledge that will be acquired for the project.
- The abilities and skills to be achieved with the project.
- Attitudes to get the students to solve the project.
- Transverse materials that will be needed by students in the project.

In addition to the above mentioned, it will also be necessary for the selected projects have some general characteristics, such as:

- The student is in the center of learning, and the project is the means to that learning. The teacher will act as facilitator of that learning.
- Students should understand the tasks to achieve the competencies identified in the curriculum as well as the importance of the project.
- Students, before starting the project, should be knowledgeable of existing products on the market and the products to be manufactured.
- The project must be the backbone of the participating subjects, as well as academics to impart knowledge.
- The methodology (POL-FC- "Kounaikenshuu") part of a real project (ideal if SL or KT), involving various subjects.
- The project will need committed and responsible students in their learning.
- The assessment will be a key component in this methodology [13].

The process to be followed in all active methodology, it should include:

- Search of more suitable materials for participants, by the teaching team projects.
- Presentation of the project by the teaching staff or participating companies. Setting goals, as:
  - Develop technical contents.
  - Skills and abilities to achieve.
  - Techniques, teaching resources used (portfolios, dates of partial deliveries, etc.).
  - Evaluation criteria.
- Training teams, by students.
- Starting the project.
- Performing various learning activities, marked in different projects.
- Project Completion.
- Critical reflection by students on what they learned in the various fields through the project.
- In parallel, teaching staff will facilitate the material content of the materials (documents, videos, references, …) for their work in the classroom [14].

It can summarize everything discussed in three stages: preparation, project development and evaluation, being this last issue, very important in the formation of students' aspects.

The evaluation will consist of several aspects to value:

- Evaluation based on performance, through individual activities.
- Results, given peer assessment, trainers-partner companies, and other students.
- Public presentations conducted peer assessment.
- Technical content taught through theoretical and practical, test examination, etc.
- Evaluation own team through: headings, test and orally, the entire learning process.

2.2 Objectives

The training objectives pursued in the described materials, are: Learn to develop, prepare, and submit a complete project, drawing Management standards, management concepts such as eco-design, sustainability, application of knowledge of other subjects and management of complementary programs.

Transversal objectives are: Training of staff and allocation of responsibilities, learning to work, being able to work as a team, improve oral and written communication, presentation techniques relying presentations, communication with the cooperation partner, if any, and/or teachers of other subjects, Finding information on the Internet, companies, NGOs, library, etc., autonomous learning, capacity for analysis and synthesis, in case of projects in other continents, basic knowledge of other cultures, to succeed in solving project proposals, basic knowledge and application of environmental technologies, eco-design and sustainability, knowledge of the principles of the theory of machines and mechanisms, knowledge and use of the principles of strength of materials, basic knowledge of production systems and manufacturing, basic knowledge and application of environmental technologies and sustainability, applied knowledge of business organization, knowledge and Skills to organize and manage projects. Knowing the organizational structure and functions of a project office, knowledge, and ability to apply graphic engineering techniques, knowledge and skills for calculation, design and testing machines, knowledge, and ability to calculate and design structures and industrial constructions.

Figure 2.1 shows some of the most relevant projects and their destination carried out by undergraduates from UC3M, UPM and CSA in the field of machines and mechanisms science (MMS) during the courses of application of the methodology described in this paper.

2.3 Methodology

The procedure for learning materials was carried out in several phases or stages. In a preliminary stage, teaching equipment Grade, discussed the marked competencies in the curriculum of different subjects, pointing out which were reached in the academic year 2013–14, and which were not achieved or were hard to get, like, for example:

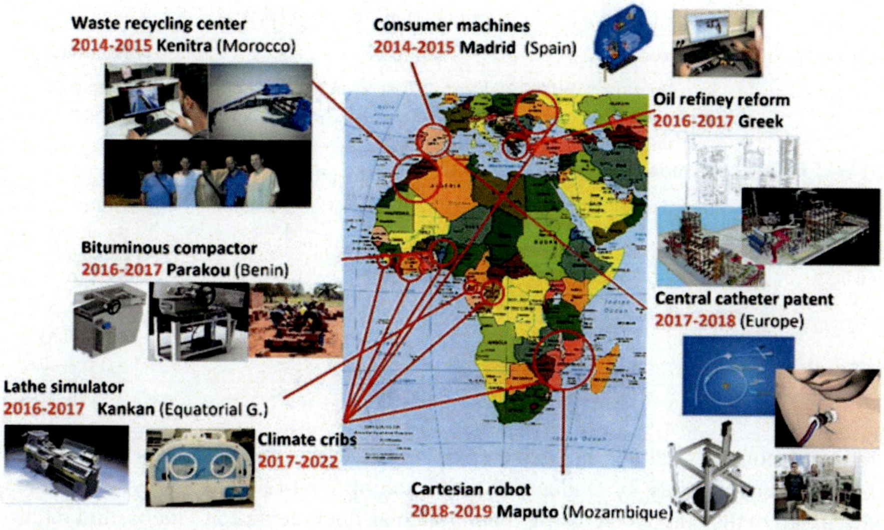

Fig. 2.1 Most relevant projects and destination

Teamwork, able to express both orally and in writing, innovation, entrepreneurship, conducting dynamic and engaging presentations, synthesis of ideas.

Once the study conducted in July 2014, a group of professors, decided to change the methodology for a more active proposal, in which students were the protagonists of their learning, through the creation of a common project subjects, with the help of their peers and the guidance and support of the faculty.

In the first phase, the teaching staff, contacted for seeking projects with businesses, municipalities, NGOs. In the end, the teaching team, opted for the International Competition Educational Innovation Projects, since the competition was in line with the contents to be taught, consisting of the design, development, calculations, plans and construction. Once the project sought to develop the teaching staff for the academic year 2014–15, used to them, learning the Japanese model "Kounaikenshuu" ("Jugyou kenkyuu" and "Jugyou Bunseki"), and for students, an active methodology hybrid, consisting of: POL and FC. During the second phase, the team of teachers, applying the methodology "Kounaikenshuu" was preparing the various teaching units; all teachers held and exchanged for improvement, the first two teaching units of different participate materials, also participating professors from CSA with the aim of: creating and improving the existing teaching materials, find new material for the network, designing questionnaires for the work of the various educational units, both at home and in the classroom, thereby obtaining more time in the classroom, for realization of the project. Once the course has begun September, students raised the new hybrid methodology, pointing out the objectives, methodology of work of the subject, skills to achieve and evaluation criteria. At first, students were not too excited because the new methodology originated a different way of working, with respect to other subjects and academic courses, involving continuous work, both at home and in

Table 2.1 Time distribution of a teaching unit (TU) in a week of class

Home	Classroom				Laboratory
Personal work	Introduction explanation teaching material	Individual test	Test team	Correction and clarification questions	Minutes project
40 to 50′	Maximum 10′	Maximum 20′	Maximum 20′	Maximum 15′	Hours week—Time project 2 h.—55′ 2.5 h.—85′ 4 h.—175′ 4.5 h.—205′

the classroom; but neither they rejected openly the idea. The third phase involved the preparation and study by students at the Teaching Unit (TU) at home, before being discussed in the classroom, using means such as documentation videos, and internet links to expand knowledge provided by the own teaching staff for this purpose. In the classroom, the weekly classes for each subject, ranging from 2 to 4.5 h per week, distributing the classroom time into 4 blocks: In the first block, the teacher makes an introduction-development of TU previously, the students have documented/prepared at home, for then students questions doubt that had arisen them during their preparation or exposure teacher, being clarified by other students or the teacher himself. The time frame for the block, ranging from five to ten minutes. Then were provided to all students, individually, a document (questionnaire) with a series of questions and reflections to respond on the TU concerned, devoting to this work an approx. time, twenty minutes. Then, once collected the individual questionnaire, were again provide the same blank questionnaire for resolution between all components of the team. In this work, they spent 20 min. Once collected the second questionnaires by professor. In Table 2.1, is showed the time distribution spent by TU along a week and the time devoted to the project according to the weekly hours per subject.

While students were debating group, the questionnaire, the teacher walked around the room, taking notes, on the attitudes of students, such as: communication, teamwork, leadership, …

As shown in Table 2.1, the preparation work, understanding and learning of the subject, is dumped outside the classroom, so that in the classroom stay longer to meet, discuss, and solve problems from traditional methodology, they raised at home.

The professors staff met once every fortnight throughout the quarter, to review, compare and correct them if needed, the process of student learning, being the fourth phase.

The fifth phase consisted, according to Japanese teaching methodology, in "Jugyou Bunseki" (lesson analysis), i.e., critical reflection, both by the teaching staff on achievements, difficulties and competencies achieved by both groups (professors and students), see Table 2.2.

Table 2.2 Outline of project work, teachers and students

Month	Week	Professors	Revision tracking professors' part	Project phases	Students	Student self-assessment working group
May–August		Search and selection of projects				
September	First	Presentation projects			Creation of working groups (3 students)	
	Second		Review underway	Project starts	M. Project Project planning	
	Fourth	First deliverable	Review underway	1st portfolio delivery Initial design booklets	Delivery and PPT presentation of achievements to date	
October	First	Valuation attitudes classroom				Rubric operating team members
	Second	Valuation attitudes classroom	Review underway			
	Third	Second deliverable		2nd portfolio delivery More advanced calculations and design	Second PPT presentation	Test operation team
	Fourth	Valuation attitudes classroom	Review underway	Communication, leadership, etc		
November	First	Third deliverable		3rd portfolio delivery Plans and budget		Rubric operating team members
	Second	Valuation attitudes classroom	Review underway		Third PPT presentation (75% done)	
	Third	Fourth deliverable		4th portfolio delivery Specifications and maintenance		Test operation team

(continued)

Table 2.2 (continued)

Month	Week	Professors	Revision tracking professors' part	Project phases	Students	Student self-assessment working group
	Fourth	Valuation attitudes classroom	Review underway			
December	First	Fifth deliverable		5th portfolio delivery Start assembly 3D printer		Rubric operating team members
	Second	Valuation attitudes classroom	Review underway		Fourth PPT presentation (90% done)	
	Third	Sixth deliverable		6th portfolio delivery 3D printer is complete Delivery		Test operation team
January	Second	Project delivery	Project final review and evaluation			
				Delivery of 3D printing machine and advertising 3D printer foam board	Final presentation of the project, critical evaluation of learning, and peer evaluation	Theory materials exam, as established date

The project was monitored through a portfolio designed for that purpose, where students, at the request of teachers, as planned previously, were making partial deliveries, considering when working, aspects such as: eco-design, sustainability, cultural criteria and technical documentation for correction and monitoring, making clear the contents made by each member of the team, ahead of their individual responsibility.

Once the project was completed, it was presented to teachers and other students, again; each group attached a report that consisted of:

- A poster in DIN A0, commenting on the fundamental characteristics of the project.
- A billboard in foam board, DIN A1 format.
- A report with the results and conclusions of the work, with particular emphasis, how the work meets the needs established the contest rules.
- A personal reflection of how the project helped them carried out, to achieve the competencies identified by the Ministry of Education as well as their entry into the professional world.

- A dossier with all documents must have a project report, State of the art, stones, Plans, Specifications, Budget, and Bibliography.

Every month students were given a survey consisting of 41 questions, like the one designed by Mike Winer, with the idea that analyzing both individual responsibility and collective teamwork towards the project was carried out.

The purpose of this tool was:

- Learn to make the necessary changes.
- Increase the efficiency and effectiveness of the working group. Focusing on how the group will celebrate the successes and will focus on meeting the challenges they still have to overcome, motivating each other to improve the harmony and effectiveness of work, and, realizing how they progress together.

They also performed monthly with a lag compared to the previous survey of fifteen days, a rubric, and thought to evaluate and analyze the performance of each member of the team evaluated by themselves, with the idea of possible problems of work. These assessment tools were a resource that the team of teachers made available to students in order that they might be the ones the discovered the strengths and weaknesses of teamwork, detecting the problems that arise in any project and try to find solutions.

The evaluation of each subject, consisted on: a theoretical exam (ET), a project (P) and the degree of student interest (I). The final grade (CF) was obtained after applying the following formula, see Eq. 2.1.

$$CF = 0.3CF + 0.65\mathrm{P} + 0.05\mathrm{I} \tag{2.1}$$

- Theoretical exam (ET). Technical knowledge of the subject taught in class.
- Project (P). Professional achievement, personal and social skills, achieved in the implementation of the POL-FC project, according to Table 2.3, to solve or handle similar projects.
- Interest (I). Evaluation by the teacher according to the interest shown by the student to the subject taught, considering: punctuality, attendance, perseverance, delivering …

The evaluation criteria of learning and its weight percentage in the project, were assigned by the teaching team by mutual agreement, you can be seen in Table 2.3.

Table 2.3 Weight of project in percentages

Proposal	Possible solution	Design	Implementation, prototypes	Oral presentation	Written documentation	Attitudes	Rating peers
5%	10%	25%	25%	10%	10%	10%	5%

2.4 Results and Analysis

It should be noted that the traditional teaching methodology, based on explaining some theoretical content, conducting practical exercises, and a review was used from the 2011–2012 academic year to 2013–14 academic year.

The final grades achieved by students of Technical Office from UC3M and UPM, and by students of Mechanical Fabrication from CSA, are in Table 2.4.

Regarding the presentations throughout the semester to present the projects, Table 2.5 shows the progression obtained from presentations made by students in the last 5 academic years.

By asking each student group each month that performed a presentation of what has been done so far, presenting its progress, it should be noted that not only improved, thanks to the instructions of teachers and peers, their presentations, but observed improvements in the synthesis of their ideas, as well as oral and written expressions.

As for the assessments made by students on their own teammates work, both in the test team, and the headings should be noted that this methodology was used to detect before they were to occur, possible problems within the team, helping, the teaching staff, to keep track of teamwork.

Table 2.4 Average rating theoretical knowledge in past 10 academic years

Methodology	Average rate UC3M (0–10)	Average rate UPM (0–10)	Average rate CSA (0–10)
No POL-FC-K course 2011–2012	5	4.9	6.2
No POL-FC-K course 2012–2013	5.3	5.3	6.35
No POL-FC-K course 2013–2014	5.4	5.1	6.3
With POL-FC-K course 2014–2015	7.2	6.5	8.1
With POL-FC-K course 2016–2017	6.9	7.1	7.9
With POL-FC-K course 2017–2018	7.3	7.2	8.2
With POL-FC-K course 2018–2019	7.6	7.4	8.4
With POL-FC-K course 2019–2020	7.1	7.5	7.8
With POL-FC-K course 2020–2021 COVID 19	7.1	7.3	7.5
With POL-FC-K course 2021–2022	7.5	7.5	8.2

Table 2.5 Average rating projects presentations in last 5 academic years

Methodology	Average rate UC3M (0–10)	Average rate UPM (0–10)	Average rate CSA (0–10)
Course 2017–2018	5.5	5.3	5.5
Course 2018–2019	6	5.9	6.2
Course 2019–2020	6.4	6.5	6.8
Course 2020–2021 COVID 19-online	7.2	7.1	7.5
Course 2021–2022	8.1	7.9	8.4

To assess the attitudes, the team of teachers had a series of tabs, like the designed by the Ministry of Education on concepts such as leadership, communication, innovation, entrepreneurship, teamwork, … being completed during business hours' project work in the classroom.

With a single investigation by this method of learning, it is not possible reach any definite conclusion, but it is possible to say that results have been different (better) than by traditional methods.

2.5 Conclusions

With this hybrid active methodology consists of: POL, FC and "Kounaikenshuu" for teachers, training has been improved in areas such as:

- Professors:
  - Increasing and improving learning faculty.
  - Improved production of teaching units, i.e. realization of didactic units that students learn issues that find it more difficult to understand.
  - Creating a framework for teamwork, achieving an improvement in the content of teaching units, while creating a sense of team.
  - Communicate and disseminate innovation generated with colleagues from the University and other universities, as is the case.

    As negative aspects can be highlighted:

    - Too much work for the faculty.
    - Poor involvement of other departments and teachers.
    - Too many students and therefore few teachers for many projects.
- Students:

  Improvements have been noted in the achievement of competencies set by the curricula of subjects, highlighting:

  - More knowledge and skills.
  - Development both oral and written.

- More ease when: plan, prepare and carry out the project.
- Management of other languages because they had to consult articles and documentation in English, French and German.
- Improved in: teamwork, leadership, innovation, entrepreneurship, research, ...

References

1. Tippelt, H., Lindemann, R.: El método de proyectos (2001)
2. Boss, S., Kraus, J.: Reinventing Project-Based Learning. Your Field Guide to Real-World Projects in the Digital Age. ISTE (2007)
3. Cárcel Carrasco, F.J.: El método de proyectos como técnica de aprendizaje en la empresa, 3C Empres. Investig. y Pensam. crítico, vol. 05, no. 01, pp. 16–28 (2016)
4. García-Alonso, J.M., Soriano-Heras, E., Blaya, F., Rubio, H.: Didactic methodologies used in industrial design and mechanical engineering for the implementation of the marked competencies and their professional insertion, vol. 64 (2019)
5. García-Alonso, J.M., Soriano, E., García-Vicario, I., Rubio, H.: Service-learning projects based on dynamic documentation in engineering colleges. In: Mechanisms and Machine Science, vol. 19 (2014)
6. Heras, E.S., Alonso, J.M.G., Haro, F.B., D'Amato, R., Marcos, E.I.M.: Metodologías didácticas utilizadas en el diseño industrial para la implementación de las competencias marcadas y su inserción profesional, Técnica Ind., vol. 320, pp. 38–44 (2019)
7. Leal Filho, W., Shiel, C., Paço, A.: Implementing and operationalising integrative approaches to sustainability in higher education: the role of project-oriented learning. J. Clean. Prod. **133**, 126–135 (2016)
8. Blaya, F., Nuere, S., Soriano, E., García, J.M., Islán, M.E.: Implementation of a methodology project oriented learning (POL) on specific subjects of bachelor and master in industrial design engineering, pp. 1352–1362 (2016)
9. Chassidim, H., Almog, D., Mark, S.: Fostering soft skills in project-oriented learning within an agile atmosphere. Eur. J. Eng. Educ. **43**(4), 638–650 (2018)
10. Kim, M.K.: Students' and teacher's reflections on project-oriented learning: a critical pedagogy for Korean ELT. English Teach. **70**(3), 73–98 (2015)
11. Bishop, J., Verleger, M.: The flipped classroom: a survey of the research. In: 2013 ASEE Annual Conference & Exposition Proceedings, pp. 23.1200.1–23.1200.18
12. Kitada, Y.: Teacher agency in the modification of Japanese lesson study in the United States. Educ. Stud. Jpn. **16**, 45–57 (2022)
13. García Peñalvo, F.J.: Libro blanco de la universidad digital 2010 (2009)
14. de Miguel Díaz, F.M.: Metodologías de enseñanza y aprendizaje para el desarrollo de competencias. Orientaciones para el profesorado universitario ante el Espacio Europeo de Educación Superior (2006)

Chapter 3
Developing Students' Cognitive Skills in MMS Study Course

E. Krylov, S. Devyaterikov, V. Balyakin, and S. Dobrovolsky

Abstract The article discusses the strategies for involving students into active learning activity during studying Mechanism and Machine Science (MMS), and developing students' cognitive competencies and metacognitive skills. Virtual labs as a learning methodologies and tools show good results and perspectives. The creative problem-based learning has been shown to be effective in education, but is not a focus of MMS courses. Brain Storming, TRIZ, Synectics, and other creative problem-solving methods can be adapted for the active MMS learning. The article suggests the adaptation of the SCAMPER, a brain storming method for solving several problems concerning structural analysis, kinematics, and gear trains.

Keywords MMS · Students · Cognitive skills · Creative thinking · Problem-based learning · SCAMPER · Heuristic method · Problem solving

The goals and objectives of engineering training are now undergoing a significant transformation. Meta-subjective cognitive skills, critical thinking, creativity become no less important than professional knowledge acquired in vocational training. In XXI century engineers face the challenges of complexity, uncertainty and ambiguity as three fundamental aspects of post-industrial technology [1]. This should give engineering education a new focus on student creativity and innovation. Rugarsia et al. note that the volume of information that engineers are called upon to know is increasing far more rapidly than the ability of engineering curricula to cover it. The solution proposed is that the focus in engineering education must shift away from the simple presentation of knowledge and toward the integration of knowledge and the development of critical skills needed to make appropriate use of it [2]. Active role of a

---

E. Krylov (✉) · S. Devyaterikov · S. Dobrovolsky
Kalashnikov Izhevsk State Technical University, Izhevsk, Russia
e-mail: 649526@mail.ru

V. Balyakin
Samara State Aerospace University, Samara, Russia

J. C. García Prada et al. (eds.), *Trends in Educational Activity in the Field of Mechanism and Machine Theory (2018–2022)*, Mechanisms and Machine Science 128,
https://doi.org/10.1007/978-3-031-25730-8_3

Table 3.1 Analytical versus contextual ways of teaching subjects

Analytical approach	Contextual approach
Students must know the fundamentals	Students must know the fundamentals
Minimal computer use	Extensive computer use
Problems are fully defined	Problems are open-ended
Students spend much time substituting in equations (plug-and-chug)	Students spend much time in critical thinking and in asking "what if" questions
Only one "correct" solution expected	Multiple solutions/alternatives expected
Pure analysis-no design content	Application to design is central
Ask good (the right) questions	Critical thinking

learner becomes very important, so cognitive engagement is cited as a critical component of an educational experience [3]. To increase cognitive engagement, students must move from shallow cognitive processing to meaningful cognitive processing [3]. Some researches note that the expectations of industry, academic and faculty are shared by students themselves: "current expectations of engineering students are not only that they have the ability to learn, to achieve and to create but also to have the ability to be self-starters, critical and creative thinkers" [4].

One way to overcome the traditional role of student as a passive receiver of content is involving him/her into the learning activity based upon virtual labs [5, 6]. This pedagogy realizes a switch from analytical approach to contextual approach in teaching subjects (see Table 3.1 [7]). Lumsdaine and Lumsdaine give arguments for contextual approach in [7].

By examining Herrmann's four-quadrant brain model of thinking together with mindsets needed for creative problem solving, Lumsdaine and Lumsdaine have shown that if the contextual approach in teaching/learning is incorporated into the engineering curriculum, then students learn much more, they can apply their skills in many new situations, they learn flexibility, leadership, and critical thinking [7].

Simulation of mechanisms is a powerful methodology to help learners better understand theory concepts, it provides them for a means of the deeper numerical analysis, and stimulates for independent learning activity. Simulation and modeling of MMS products contribute greatly in students' comprehension of kinematics and dynamics of mechanisms. In their recent paper Ceccarelli and Cocconcelli presented an historical analysis of development of mechanism models used in academic teaching fields, and revealed the didactic potential of CAD models to analyze different kinematic behaviors of mechanisms [8].

From didactic point of view it is important to let the learners feel the designing as an open-ended process: how any change in the input (configuration, number of links, types of kinematical pairs, loads) will affect the final structure of the mechanism and its parameters. Open-ended problems methodology needs appropriate methods and software tools to provide students with an instrument that is ready-to-use, user-friendly, and easy-to-change when applied for different mechanisms. The method of vector closed contours considers a mechanism as a combination of planar or spatial elementary vector modules, depending on the set of linear and angular arguments [9]. For a given structural scheme students develop the parametric formula for the mechanism's vector model and use specialized KDAM software for numerical analysis [10, 11]. KDAM provides a means of easy-to-use investigation of a mechanism at the first stages of design and the optimization of its key parameters (dimensions, pressure angles, reduced loads, masses, reactions in joints, etc.).

Problem-based learning (PBL) and project learning have been proved to be effective in facilitating students' development of higher order learning and skills. PBL has become increasingly popular in K-12 and higher education worldwide since it was first introduced in medical education in the late 1960s [12]. However, even in XXI century it has not gained significant popularity in engineering curricula due to the large time-scale needed to solve complex engineering problems and the difficulties associated with assessment of its impact on students [13]. There is even some criticism of the benefits of widespread adoption of this method. J. Perrenet et al. claim that PBL has certain limitations, which make it less suitable as an overall strategy for engineering education [12].

Mechanism and Machine Science (MMS) as a study course has much in common with problem-based learning, since it also can be considered as a problem-centered science. Any piece of theory should be illustrated with examples, and many problems come from past or modern industry. So, solving problems in regular course of MMS should also hold promise for cultivating students' creativity. This is especially important for training students for MMS competitions—Olympiads [14, 15]. Very often a contest problem does not have overcomplicated solution, but participant should be experienced thinker and creative person to identify this way of solution and follow it until the problem is solved.

Brain Storming, TRIZ, and Synectics are creative problem-solving methods that can be adapted for engineering design courses. These and other methods use algorithms based on principles, techniques, and operators. Some of them are intuitive, some more analytical, but all they are heuristic. The common milestones of problem solving are: define the problem, analyze causes, generate ideas, weigh up ideas, make a decision, determine next steps to implement the solution, evaluate whether the problem was solved or not.

The aim of this article is to demonstrate the applicability of some heuristic techniques for creative solving MMS problems. In this respect, the phase of idea generating is especially interested. It can be organized with SCAMPER, the brainstorming methods using a set of directed idea-spurring questions. The questions inspire changes in thinking process, give rise to a new vision of the problem.

The changes that SCAMPER stands for are: S—Substitute; C—Combine; A—Adapt; M—Magnify/Modify; P—Put to other uses; E—Eliminate; R—Rearrange/Reverse [16]. Here we'll demonstrate the application of three heuristic solution methods for MMS problems. These methods, which correlate with the elements of the SCAMPER spectrum, have proved to be effective for use in the MMS study course.

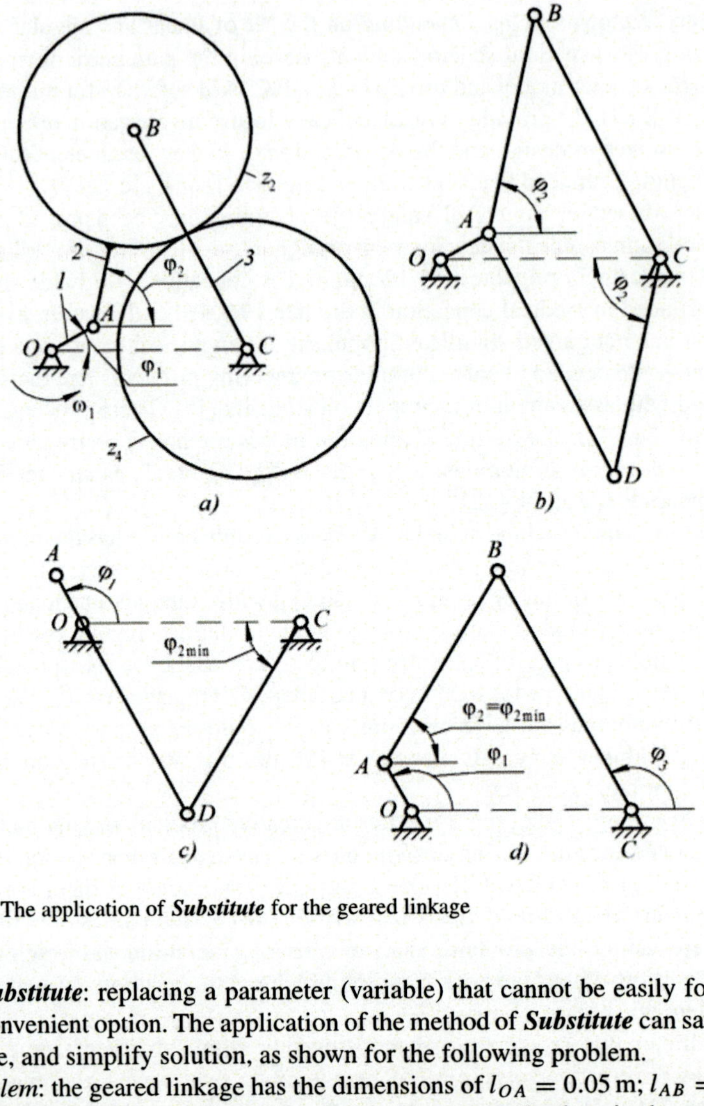

Fig. 3.1 The application of ***Substitute*** for the geared linkage

1. ***Substitute***: replacing a parameter (variable) that cannot be easily found by a more convenient option. The application of the method of ***Substitute*** can save efforts and time, and simplify solution, as shown for the following problem.

Problem: the geared linkage has the dimensions of $l_{OA} = 0.05$ m; $l_{AB} = 0.20$ m; $l_{BC} = 0.25$ m; $l_{OC} = 0.20$ m (Fig. 3.1a). The gear wheel z_2 is the part of the connecting rod *AB*, while the wheel z_4 rotates about a fixed axis passing through the center of the hinge *C*. The numbers of teeth are: $z_2 = 25$; $z_4 = 35$. Crank *OA* rotates uniformly with an angular velocity of $\omega_1 = 70$ rad/s.

For the special position where angle φ_2 takes the minimum value (i.e., $\varphi_2 = \varphi_{2_{min}}$), *find*: (*i*) value of angle φ_1; ; (*ii*) value of angle $\varphi_{2_{min}}$; ; (*iii*) angular speed of gear wheel z_4.

It can be easily seen that *OABC* is a crank and rocker mechanism, hence functions $\varphi_2 = \varphi_2(\varphi_1)$ and $\varphi_2 = \varphi_2(t)$ are continuous ones. Thus, for the position in question, $\frac{d\varphi_2}{dt} = \omega_2 = 0$.

But this means that link 2 instantaneously translates, and all its points move with the same velocity. In particular, $V_B = V_A$, from this it follows that

$$\omega_3 = \frac{V_B}{l_BC} = \frac{V_A}{l_BC} = \frac{\omega_1 l_O A}{l_BC} = 14\ \text{s}^{-1}.$$

Gears z_2 and z_4 form an epicyclic gear chain with handle *BC*. Then, angular speed of gear z_4 is found by the formula Willis's method:

$$\frac{\omega_2 - \omega_3}{\omega_4 - \omega_3} = -\frac{z_4}{z_2} = i_{24}^{BC}, \qquad \omega_4 = \omega_3 + \frac{\omega_2 - \omega_3}{i_{24}^{BC}}.$$

Now about value of angle $\varphi_{2_{min}}$: it is difficult to find it in the original linkage. The simplification is possible with the method of ***Substitute***: we introduce an imaginary linkage *OADC* provided that $l_{CD} = l_{AB}, l_{AD} = l_{BC}$. For this new linkage it is easily seen that $\varphi_2 = \varphi_{2_{min}}$ when $OD = OD_{min} = l_{AD} - l_{OA} = 0.20\,\text{m}$ (Fig. 3.1b).

Because of data given, it happens that $OD_{\text{min}} = l_{OC} = l_{CD}$, so triangle *OCD* is the equilateral one, and all the desired angles are found immediately (Fig. 3.1c):

$$\varphi_2 = \varphi_{2_{\text{min}}} = 60°; \quad \varphi_1 = \varphi_3 = 120°.$$

Figure 3.1d illustrates the position of the four bar linkage that corresponds to the value of $\varphi_{2_{min}}$. It is obvious that angular speeds of rocker (ω_3) and crank (ω_1) are of the same instantaneous direction, hence using the above formula we finally get

$$\omega_4 = 14 + \frac{(0 - 14)}{-35/25} = 24\ \text{s}^{-1}.$$

2. ***Modify***: Can you change the item in some way? Can you start your solution with something not completely known?

Very often the solution of a coplanar MMS problem is found from the vector polygon. In some cases the order of components in a sequence of vectors does matter, and reasonable choice leads to significant simplifications. Sometimes the initial chain of the polygon is unknown in magnitude; nevertheless solution can be obtained by combining this chain with others.

Kinematical analysis provides good examples for the application of the method of ***Modify***. Consider the following problem. In a coplanar mechanism (Fig. 3.2) links 1 and 2 are connected by rod 2 and pin-in-the-slot unit at *E*. The data given are: $\omega_1 = 2\,\text{rad/s}$, $O_1A = O_1E = ED = BD = 100\,\text{mm}$, O_2D=150 mm.

It is asked for angular speed ω_4 at the position given in Fig. 3.2.

There are two possible geometrical methods of solution: instant centers and vector polygon. Here we'll use the second one.

The expressions for velocities of point B with respect to points A and D: $\vec{V}_B = \vec{V}_A + \vec{V}_{BA} = \vec{V}_D + \vec{V}_{BD}$, where $\vec{V}_A \perp O_1A$; $\vec{V}_{BA} \perp AB$; $\vec{V}_D \perp O_2D$; $\vec{V}_{BD} \perp BD$.

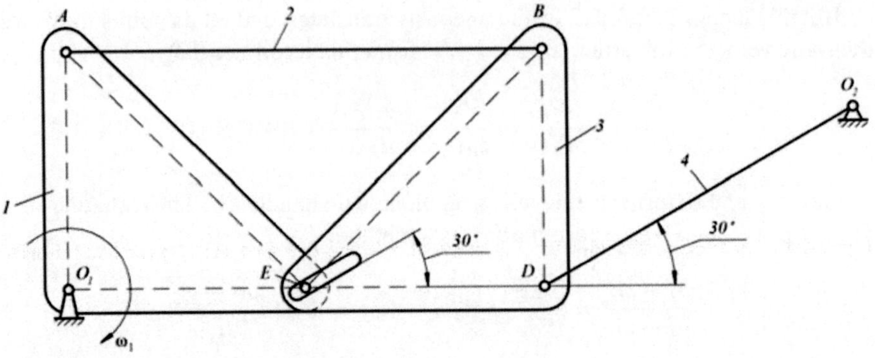

Fig. 3.2 The application of *Modify* for planar kinematics

Fig. 3.3 Velocity polygon

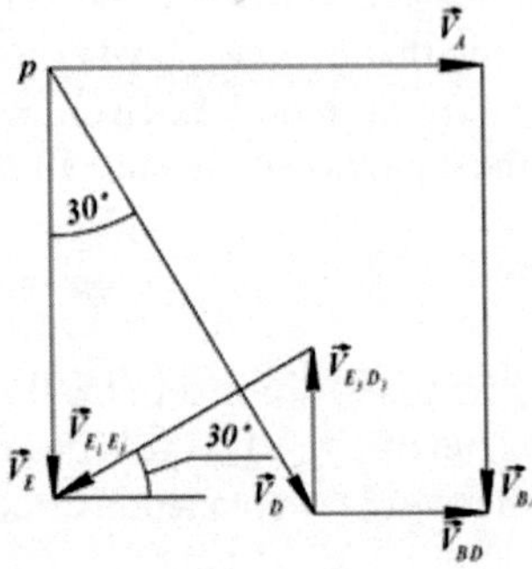

Since this equation contains only one value, known completely (V_A), that is not possible to find solution immediately. But one can make a guess: vector polygon starts with $\overrightarrow{V}_D$ *that is laid off to an arbitrary scale* (Fig. 3.3). We also lay off $V_A = V_E$ as rays from the same origin, taking their directions into account.

Finding velocity of point E makes the solution closed:

$$\vec{V}_{E_1} = \vec{V}_{E_3D_3} + \vec{V}_{E_1E_3}, \quad \frac{V_{E_3D_3}}{V_{BD}} = \frac{ED}{BD} = 1.$$

The tip points I, II, and III serves as the reference points in the velocity diagram. By drawing one line through every of them, in according to the equations above, one can close the polygon (Fig. 3.3). The readings in the picture are:

$$V_A \to x;\ V_E \to x;\ V_{BA} \to a;\ V_{BD} = V_{E_3D_3} \to b;\ V_D \to d;\ V_{E_1E_3} \to c.$$

From the polygon we have

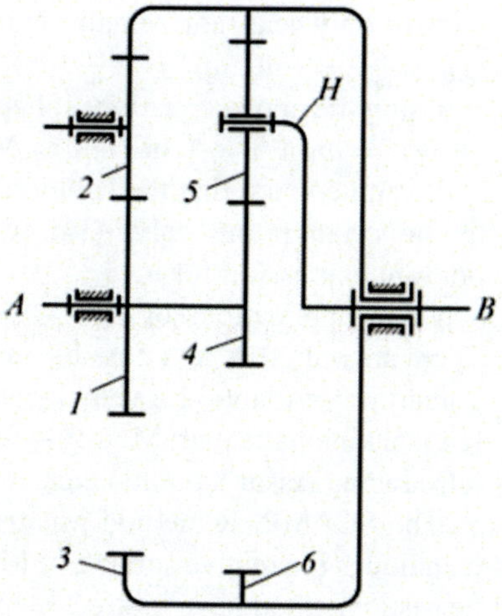

Fig. 3.4 The application of ***Eliminate*** for epicyclic gear train

$$\begin{cases} d\sin 30^\circ = x - b \\ d\cos 30^\circ = x + b - c\sin 30^\circ \\ d\sin 30^\circ = c\cos 30^\circ \end{cases}.$$

The answer: $d = 1.2x$, or $V_D = 1.2(\omega_1 \cdot O_1A) = 240\,\text{mm/s}$, and $\omega_4 = \frac{V_D}{O_2D} = 1.6\,\text{s}^{-1}$.

3. ***Eliminate, Reduce***: What unnecessary issues can you eliminate? Focus on question strictly. Avoid actions that are not required and that are not absolutely necessary.

Many MMS problems are complicated for both reasoning and computation. Sometimes it seems that few values are missing in the data given. What to do, how to solve? Students can become frustrated and upset, or even fall into a stupor. One can remedy the situation by focusing only on relevant piece of big puzzle, letting irrelevant information float away freely.

Often kinematical analysis of a gear box demands the answer for angular speed of every gear wheel. Yet in the problem below the question is reduced to something less general one.

Problem. The input shaft A in the gear box (Fig. 3.4) makes $N_A = 1440$ rpm. The gear ratio is $i_{AB} = \frac{\omega_A}{\omega_B} = -40$, the number of teeth $z_4 = z_5$.

Find N_{5H}, the relative velocity of gear z_5 with respect to handle H.

From the beginning one should focus closely on the key word: *relative velocity*, the angular velocity of gear 5 as seen from the handle H. It is

$$\omega_{5H} = \omega_5 - \omega_H.$$

The formula (Willis') method for the gears 4 and 5 provides $\frac{\omega_4 - \omega_H}{\omega_5 - \omega_H} = -\frac{z_5}{z_4}$, but $z_5 = z_4$, and $\omega_4 = \omega_1$ for the mechanism.

Hence $\omega_1 - \omega_H = \omega_5 - \omega_H = \omega_{5H}$.

Now only gear ratio remains unused, the ratio of input and output gear velocities: $i_{AB} = \frac{\omega_1}{\omega_H} = -40, \omega_H = -\frac{\omega_1}{40}$.

Using the above equations it happens that relative velocity is equal to $\omega_{5H} = \omega_1(1 + \frac{1}{40})$, or $N_{5H} = N_1\left(1 + \frac{1}{40}\right) = 1440 \cdot \frac{41}{40} = 1476$ rpm.

It can be noted that the solution was pretty simple, because we were focused on the important points only. Also, we needed clear idea about relative speed and the relevant expression for it.

Other applications of SCAMPER technique can be found in MMS problems. There are only few next possible ones: Rearrange—for closure of a force polygon in a smart way; Adapt—for velocity and acceleration analysis of the links making translational kinematic pair; Modify—for velocity analysis of 3-d-class Assur's group (by introducing Assur's points) etc.

The SCAMPER method was used as a sample in order to think about heuristic techniques in creative solving of MMS problems, so highly likely the specific MMS heuristic method have indirect mapping with SCAMPER components.

Conclusion

Engineering faces many challenges and controversies, the number of which has increased significantly in recent decades. To be successful in the profession a student should gain cognitive competencies and metacognitive skills during studying in university. Engineering curriculum can provide the development of these personal qualities, including creative and critical thinking, if special types of activities are incorporated in study subjects. In engineering pedagogy a switch from pure analytical approach to contextual approach can be realized in two modes: intensive use of computers in learning process, and creative problem-based learning. MMS as one of core engineering sciences provides many opportunities for both. Virtual labs as a learning methodologies and tools are increasingly being used by MMS educators. But it seems that problem-based learning is not in the focus of discussion. Brain Storming, TRIZ, Synectics, and other creative problem-solving methods can be adapted for MMS courses. This article shows the adaptation of the SCAMPER method for solving several problems concerning structural analysis, kinematics, and gear trains. Authors hope that the creative problem-based learning, regardless of the specific platform on which it is implemented, will make a significant contribution to the cognitive development of future mechanical engineers.

References

1. Kastenberg, W.E., Hauser-Kastenberg, G., Norris, D.: An approach to undergraduate engineering education for the 21st century. In: 36th ASEE/IEEE Frontiers in Education Conference, San Diego, CA, Institute of Electrical and Electronics Engineers (IEEE), pp. 1497–1502 (2006)
2. Rugarcia, A., Felder, R., Woods, D.R., Stice, J.E.: The future of engineering education: Part 1. A vision for a new century. Chem. Eng. Educ. 34, 16–25 (2000)

3. Barlow, A., Brown, S., Lutz, B., et al.: Development of the student course cognitive engagement instrument (SCCEI) for college engineering courses. IJ STEM Ed **7**, 22 (2020). https://doi.org/10.1186/s40594-020-00220-9
4. Rao, K., Nuggenahalli, N., Ashwini, B.: Emphasis on the cognitive framework in teaching—learning process in engineering education: an empirical overview. J. Eng. Educ. Transform. 175, 175–181 (2015). https://doi.org/10.16920/ijerit/2015/v0i0/59354
5. Macho, E., Urízar, M., Petuya, V., Hernández, A.: Improving skills in mechanism and machine science using GIM software. Appl. Sci. 11, 7850 (2021). https://doi.org/10.3390/app11177850
6. Suñer, J.L., Carballeira, J.: Enhancing mechanism and machine science learning by creating virtual labs with ADAMS. In: García-Prada, J.C., Castejón, C. (eds.) New Trends in Educational Activity in the Field of Mechanism and Machine Theory, pp. 221–228. Springer, Berlin/Heidelberg, Germany (2014)
7. Lumsdaine, M., Lumsdaine, E.: Thinking preferences of engineering students: implications for curriculum restructuring. J. Eng. Educ. 84 (1995). https://doi.org/10.1002/j.2168-9830.1995.tb00166.x
8. Ceccarelli, M., Cocconcelli, M.: Italian historical developments of teaching and museum valorization of mechanism models. Machines 10, 628 (2022). https://doi.org/10.3390/machines10080628
9. Kosenok, B., Balyakin, V., Krylov, E.: Method of closed vector contours for teaching/learning MMS. In: García-Prada, J., Castejón, C. (eds.) New Trends in Educational Activity in the Field of Mechanism and Machine Theory. Mechanisms and Machine Science, vol. 64, pp. 3–10. Springer, Berlin/Heidelberg, Germany (2019)
10. Kosenok, B., Balyakin, V., Krylov, E.: Dimensional synthesis of a cam profile using the method of closed vector contours in the theory of machine and mechanism study course. In: Mechanisms and Machine Science (book series), vol. 73. pp. 753–763 (2019). https://doi.org/10.1007/978-3-030-20131-9_75
11. Kosenok, B., Balyakin, V., Krylov, E.: Method of vector closed contours in design problems of study course "Internal combustion engines: kinematics and dynamics". In: Mechanisms and Machine Science (book series). 2019. Vol. 73. Pp. 775–784. – DOI https://doi.org/10.1007/978-3-030-20131-9_77.
12. Perrenet, J.C., Bouhuijis, P.A., Smits, J.: The suitability of problem-based learning for engineering education: theory and practice. Teach. High. Educ. 5(33), 345–358 (2000)
13. Hunt, E., Lockwood-Cooke, P., Kelley, J. Linked-class problem-based learning in engineering: method and evaluation. Am. J. Eng. Educ. 1(1), 79–87 (2010)
14. Balyakin, V., Krylov, E.: Cultural and educational significance of MMS competitions for future engineers. In: Mechanisms and Machine Science (book series), vol. 64, pp. 3–10 (2019).https://doi.org/10.1007/978-3-030-00108-7_5
15. Krylov, E.G., Devyaterikov, S.A., Gubert, A.V., Egorova, O.V.: SIOMMS: evolution and development. Mech. Mach. Theory 153, 104029 (2020). https://doi.org/10.1016/j.mechmachtheory.2020.104029.
16. Serrat, O.: The SCAMPER technique (2009). https://www.researchgate.net/publication/239823670_The_SCAMPER_Technique

Chapter 4
Kinematical Analysis and Dimensional Synthesis of *RRPR-Type* Four-Bar Mechanism in MMS Study Course

S. Devyaterikov, E. Krylov, A. Gubert, and A. Nazarov

Abstract The article discusses the modification of RRPR linkage containing D31 dyad. The analysis shows that the angular position of the leg with respect to the slider has no effect on the motion of the output link. The chains with different angles between leg and slider are kinematically isomorphic and equivalent to the basic mechanism with a right angle between these elements. The investigation of limiting positions allows finding simple relationships that predict the rotational behavior of this mechanism modification based on the link lengths. Such an analysis can contribute to the MMS Olympiads process and can be used in MMS study courses. It fills in the gaps and builds bridges between the different parts of the study course regarding the popular simple plane mechanism.

Keywords MMS · Four-bar linkage · D31 dyad · Dimensional synthesis · Rotability of driver link · Limiting positions

Plane four-bar linkage is one of simplest and well investigated mechanisms. Yet, because of variety of applications its structure, dimensional synthesis, and kinematical analysis of the four-bar are included in every MMS course. Besides, the research interest towards the mechanism has not been lost. Goessner specifies a valid initial pose that is important for simulating four-bar linkages [1]. Here we focus on RRPR type of the mechanism only. One can connect a dyad D31 (Fig. 4.1) to a driver link to create a coulisse mechanism with one DOF [2–5].

As a rule, structural analysis deals with the modification of D31 dyad equipped with a leg attached to the translating member. But the problems of dimensional synthesis and kinematical investigation are usually solved for different modification of the dyad, containing no legs (Fig. 4.2a). As a result some issues escape analysis, namely how length b (Fig. 4.2b) and the angle between links 2 and 3 (Fig. 4.2c) affect kinematics of the mechanism and rotability of the driver link 1.

S. Devyaterikov · E. Krylov (✉) · A. Gubert · A. Nazarov
Kalashnikov Izhevsk State Technical University, Izhevsk, Russia
e-mail: 649526@mail.ru

J. C. García Prada et al. (eds.), *Trends in Educational Activity in the Field of Mechanism and Machine Theory (2018–2022)*, Mechanisms and Machine Science 128,
https://doi.org/10.1007/978-3-031-25730-8_4

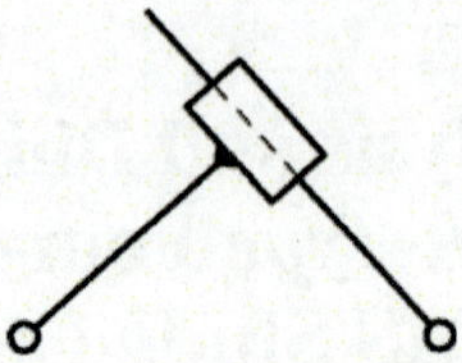

Fig. 4.1 D31 dyad

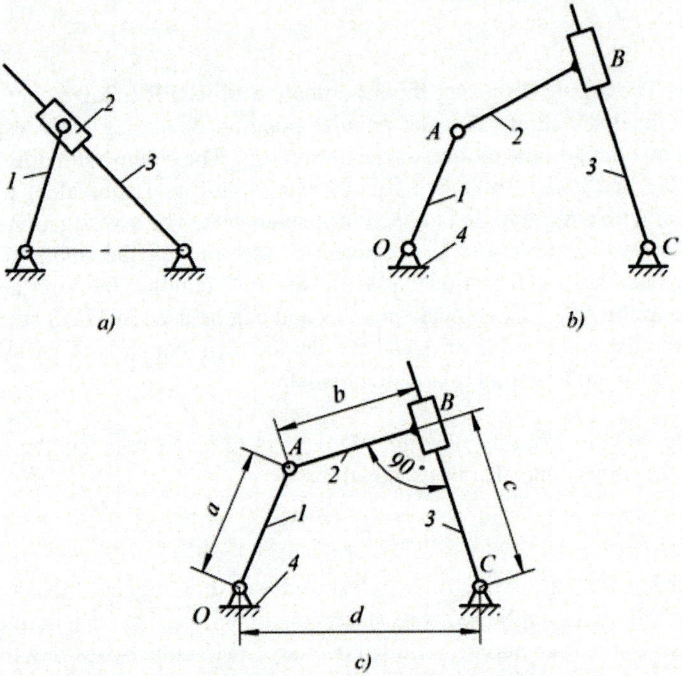

Fig. 4.2 Modifications of four-bar coulisse mechanism

Because of prismatic kinematical pair and translatory relative motion between links 2 and 3, the actual position of point D (and the value of angle between connecting bar 2 and coulisse 3) has no effect on the velocity and acceleration of the output link 3. For the mechanism (Fig. 4.3) this can be found from velocity (Fig. 4.4a and b) and acceleration diagrams (Fig. 4.5b and c). The corresponding vector equations are:

$$\vec{v}_A + \vec{v}_{BA} = \vec{v}_{B_3} + \vec{v}_{B_2B_3},$$

$$\vec{v}_A + \vec{v}_{DA} = \vec{v}_{D_3} + \vec{v}_{D_2D_3}, \text{ where } \vec{v}_{D_2D_3} = \vec{v}_{B_2B_3}.$$

$$\vec{a}_A + \vec{a}^n_{BA} + \vec{a}^t_{BA} = \vec{a}^n_{B_3} + \vec{a}^t_{B_3} + \vec{a}_{B_2B_3} + \vec{a}^k_{B_3},$$

$$\vec{a}_A + \vec{a}^n_{DA} + \vec{a}^t_{DA} = \vec{a}_{D_3} + \vec{a}_{D_2D_3} + \vec{a}^k_{D_2D_3},$$

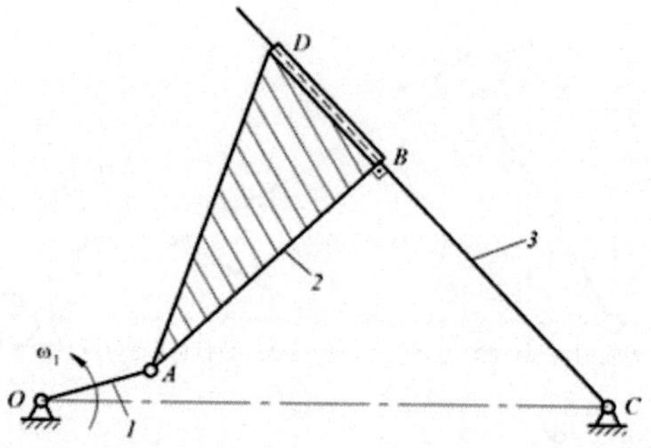

Fig. 4.3 The mechanism with extended link 2

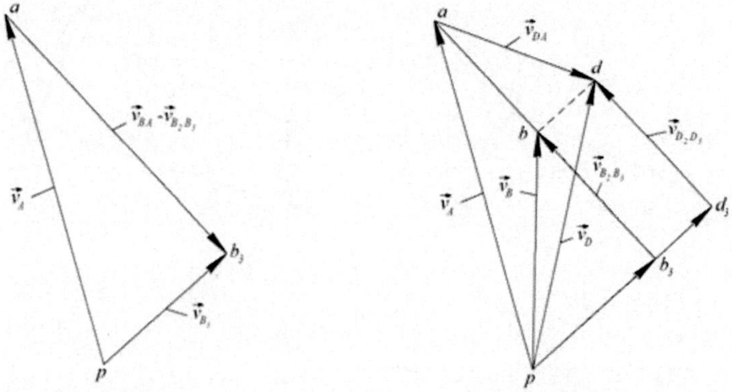

a) point *B* *b)* point *D*

Fig. 4.4 Velocity diagrams

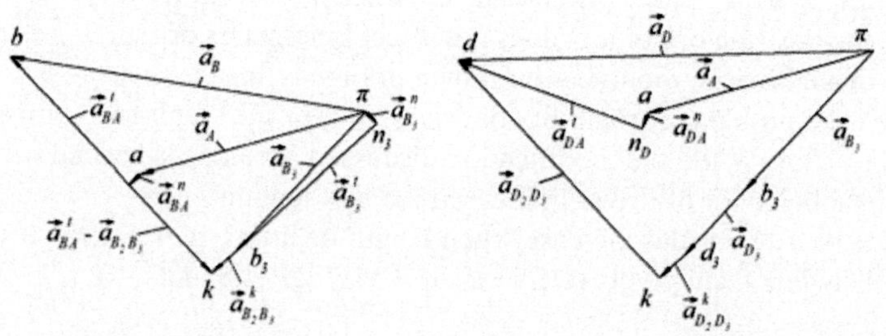

a) point *B* *b)* point *D*

Fig. 4.5 Acceleration diagrams

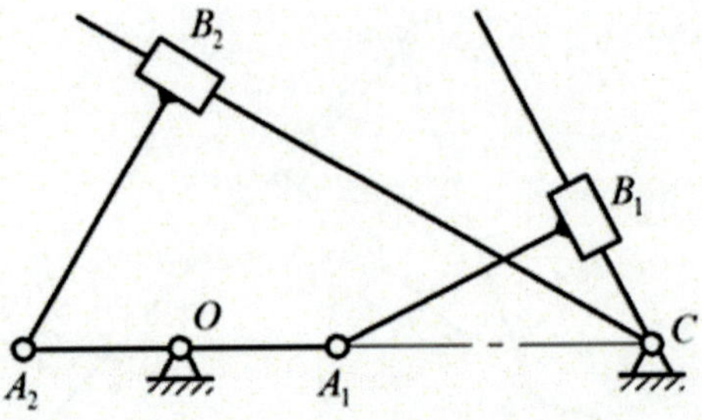

Fig. 4.6 Driver's rotability for $a < d$

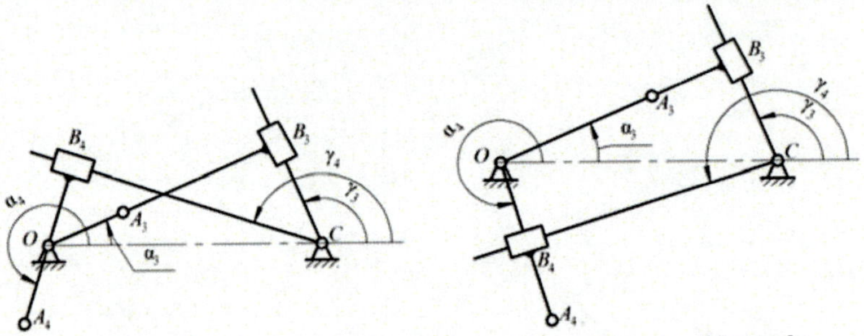

a) $a < b$

b) $a > b$

Fig. 4.7 Limiting positions for crank-and-rocking coulisse mechanism

where $\vec{a}_{D_2D_3} = \vec{a}_{B_2B_3}$, and Coriolis acceleration $\vec{a}^k_{D_2D_3} = \vec{a}^k_{B_2B_3}$.

Since the position of point D does not affect kinematics of link 3, it is advised to students to analyze the modification shown in Fig. 4.2b.

There is no problem of rotability of driver link for the crank and coulisse mechanism (Fig. 4.2a). While the modification discussed here needs special investigation of rotational behavior of driver link based the link lengths.

Case $a < d$**.** Consider the case when length of link 1 ($OA = a$) is less than the distance between fixed pivots ($OC = d$), $a < d$ (Figs. 4.6 and 4.7).

$$\text{From triangle } A_1B_1C\text{: } A_1C > A_1B_1 \text{ or } d - a > b. \tag{4.1}$$

$$\text{From triangle } A_2B_2C\text{: } A_2C > A_2B_2 \text{ or } d + a > b. \tag{4.2}$$

If the inequality (4.1) is true, then the inequality (4.2) is true for any sizes a, b, and d. Hence, the condition of rotability of driver link 1 is found from expression (4.1):

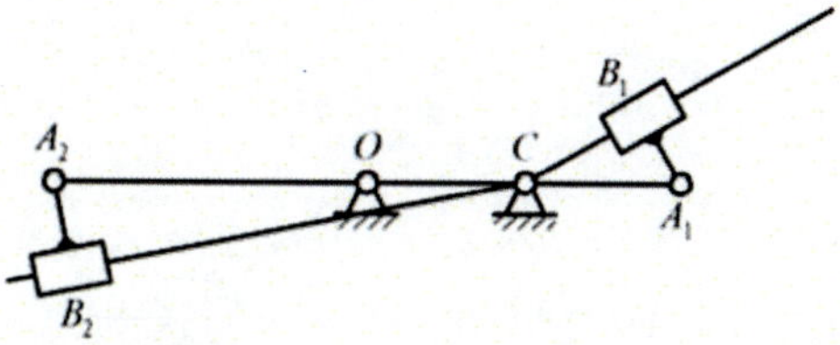

Fig. 4.8 Rotability of driver link provided $a > d$

$$a + b < d.$$

Under this condition driver link 1 is a *crank*, and we get *crank-and-rocking coulisse mechanism.* The amplitude of rocking motion is found from the consideration of the mechanism limiting positions. These positions for $a < b$ are illustrated in Fig. 4.7a.

In Fig. 4.7a the angle α_3 corresponds to the coulisse angle $\gamma_3 = \gamma_{min}$. From triangle OB_3C: $\alpha_3 = \arccos\frac{b+a}{d}$, and $\gamma_3 = \gamma_{min} = 90° + \alpha_3$.

From triangle OB_4C: $\alpha_4 = 180° + \arccos\frac{b-a}{d}$, and $\gamma_4 = \gamma_{max} = 90° + \arccos\frac{b-a}{d} = \alpha_4 - 90°$.

Then, the angle of the coulisse swivel is found as $\gamma_{max} - \gamma_{min} = \arccos\frac{b-a}{d} - \arccos\frac{b+a}{d} = \alpha_4 - \alpha_3 - 180°$.

A similar study could be carried out for the modification with $a > b$ (Fig. 4.7b).

In both mechanisms illustrated in Fig. 4.7, increasing dimension $a + b$ leads to the bigger swivel angle of coulisse because of reduction in the angle $\gamma_3 = \gamma_{min}$. If a, the length of link 1 increases, provided $a + b = const$, then both $\gamma_3 = \gamma_{min}$, and the angle of swivel are getting bigger.

Case $a > d$. Figure 4.8 shows two mechanism positions such that in the both link 1 is parallel to a fixed link 4.

$$\text{From triangle } A_1B_1C : A_1C > A_1B_1, \text{ or } a - d > b. \tag{4.3}$$

$$\text{From triangle } A_2B_2C : A_2C > A_2B_2, \text{ or } a + d > b. \tag{4.4}$$

If the inequality (4.3) is true, then the inequality (4.4) is satisfied for any dimensions of $a, b,$ *and* d. Whence, from (4.3) the condition of full rotation of driver link 1 is stated as.

$$a - b > d. \tag{4.5}$$

This is a crank-and-coulisse mechanism with a rotating coulisse.

If conditions (4.3) or (4.5) are not satisfied, then driver link 1 is not allowed to make a full revolution, and the mechanism is of a rocker-and-coulisse type.

In this case, if $a + b > d$ and $a < d$ (Fig. 4.9a), then the minimum and maximum angles of rotation of driver rocker 1 and coulisse 3 are found as follows:

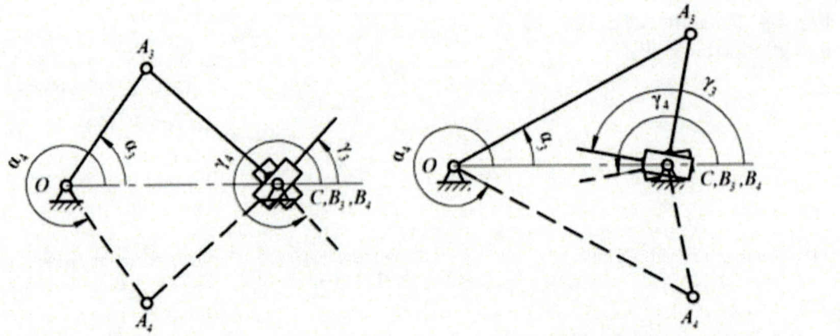

a) $a < d$ b) $a > d$

Fig. 4.9 Rocker-and-coulisse mechanism

Fig. 4.10 Vector contour

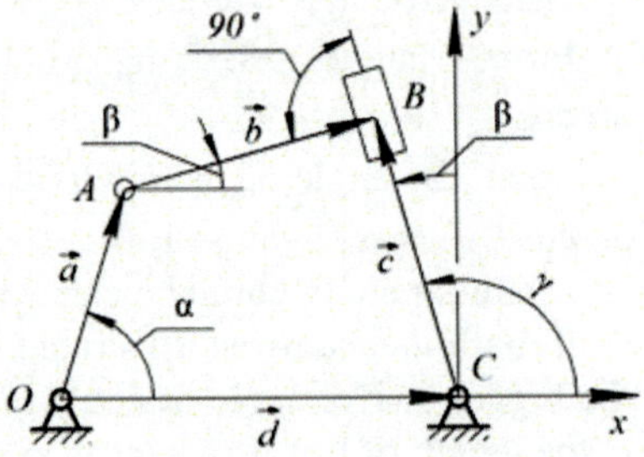

$$\alpha_{min} = \alpha_3 = \arccos\frac{a^2 - b^2 + d^2}{2ad}, \quad \alpha_{max} = \alpha_4 = 360^\circ - \alpha_3,$$

$$\gamma_{min} = \gamma_3 = 90^\circ - \arccos\frac{-a^2 + b^2 + d^2}{2bd}, \quad \gamma_{max} = \gamma_4 = 360^\circ - \gamma_3.$$

For the case $a - b < d$ and $a > d$ (Fig. 4.9b), the extreme angles are:

$$\alpha_{min} = \alpha_3 = \arccos\frac{a^2 - b^2 + d^2}{2ad}, \alpha_{max} = \alpha_4 = 360^\circ - \alpha_3,$$

$$\gamma_{min} = \gamma_3 = 270^\circ - \arccos\frac{-a^2 + b^2 + d^2}{2bd},$$

$$\gamma_{max} = \gamma_4 = 90^\circ + \arccos\frac{-a^2 + b^2 + d^2}{2bd}.$$

Dimensional synthesis of a four-bar linkage is well done by the method of vector contours [5, 6]. For the modification under discussion the contour reads $\vec{a}+\vec{b} = \vec{d}+\vec{c}$ (Fig. 4.10).

The projections are:

$$x{:}\, a\cos\alpha + b\cos\beta = d + c\cdot\cos\gamma;$$
$$y{:}\, a\sin\alpha + b\sin\beta = c\cdot\sin\gamma.$$

After some manipulations and using the condition $\gamma = 90° + \beta$, one can get:

$$\frac{a}{d}\sin(\gamma-\alpha) + \frac{b}{d} = \sin\gamma.$$

Introducing $\frac{a}{d} = k_1$ и $\frac{b}{d} = k_2$, finally

$$k_1\sin(\gamma-\alpha) + k_2 = \sin\gamma.$$

So, the dimensional synthesis of the mechanism can be performed on the base of only two given positions of crank and coulisse.

Thus, for given angular positions of link *1* (α_1, α_2) and corresponding positions of coulisse *3* (γ_1, γ_2),

$$k_1 = \frac{\sin\gamma_1 - \sin\gamma_2}{\sin(\gamma_1-\alpha_1) - \sin(\gamma_2-\alpha_2)},$$
$$k_2 = \frac{\sin(\gamma_1-\alpha_1)\sin\gamma_2 - \sin(\gamma_2-\alpha_2)\sin\gamma_1}{\sin(\gamma_1-\alpha_1) - \sin(\gamma_2-\alpha_2)}.$$

Choosing the value of distance d between fixed pivots will get desired lengths of the links $a = k_1 d$, $b = k_2 d$. Next operation should be inspection of the mechanism for rotability of driver link 1.

For example, if $\alpha_1 = 105°$, $\alpha_2 = 280°$, $\gamma_1 = 140°$, and $\gamma_2 = 160°$, then $k_1 = 0.209$, $k_2 = 0.523$. For $d = 360$ mm, the link lengths are $a = 75$ mm, $b = 188$ mm. Since $a < d$ and $a + b < d$, the mechanism is of crank-and-rocking coulisse (Fig. 4.7a).

The authors hope that the research will find applications in the following branches of MMS study course.

1. ***Dimensional synthesis of RRPR-type four-bar linkages***. A very important consideration when synthesizing a mechanism is to ensure that the input link can make a complete revolution. For RRRR planar linkage the problem is solved with Grashof's law. This is not a problem for the quick return mechanism of RRPR structure shown in Fig. 4.2c. However, for the modification containing D31—dyad the ratability problem is again important and not trivial. So, students are given the opportunity to discuss and solve non-trivial synthesis problem by themselves. This is especially important for the training students in the process of MMS Olympiads.
2. ***Kinematical analysis.*** Students' skills in kinematics can be improved because of understanding a kind of kinematical isomorphism of connecting bars forming the dyad D31. Very often the kinematical analysis of the links joined by a prismatic kinematical pair is reduced to discovering the fact that the links have equal angular

velocities because of no relative rotation. The shape of the links is usually beyond the analysis, so students lose the way of solving if connecting bar does possess any shape differ that shown in Fig. 4.1. The research in this paper shows that the shape of connecting bar (in particular, the angular position of the leg with respect to slider) does not affect the velocity and acceleration calculation. Hence, the kinematical analysis of a "family" of D31—dyads can be reduced to a basic mechanism with the leg attached to the slider at a right angle.

The listed applications show the use of the research for *educational purposes*. Besides, different parts of the MMS study course are sometimes inconsistent: mechanisms suitable for structural synthesis analysis are not addressed in the kinematics, and vice versa. For example, D31—dyads are rather popular in structural analysis but never being under consideration in kinematics. We recommend focusing on complete analysis of the mechanisms and their parts to fill the gap between MMS study course and provide students with a chance of thorough acquisition of knowledge, skills and abilities. This way of teaching follows *didactic principle* of systematization and continuity and increases students' comprehension and motivation.

Conclusion

Four-bar linkages of RRRR, RRRP, and RRPR structures are considered in all parts of a MMS study course, including structural analysis, dimensional synthesis, and kinematical & dynamical analysis. Yet the modification of RRPR mechanism with a special shape of the sliding member equipped with a leg, is considered mainly in the context of a structure. The analysis shows that the angular position of the leg with respect to the slider has no effect on the motion of the output link. All such chains are kinematically isomorphic and equivalent to the basic mechanism with a right angle between the leg and the slider. The investigation of limiting positions allows to find simple relationships that predict the rotational behavior of the mechanism modification based on the link lengths. The authors hope that such an analysis can contribute to the MMS Olympiads process [7], overall activity of IFToMM [8], and will be used in MMS study courses. It probably fills in the gaps and builds bridges between the different parts of the study course regarding the popular simple plane mechanism.

References

1. Goessner, S.: Valid initial pose of any fourbar linkage. https://www.researchgate.net/publication/319256978_Valid_Initial_Pose_of_Any_Fourbar_Linkage (2017).
2. Marghitu, D.: Kinematic Chains and Machine Components Design. Elsevier Academic Press, 778 p. (2005)
3. Artobolevsky, I.I.: Theory of Mechanisms and Machines, 4th edn. Science, 639 p. (2009)
4. Khurmi, R.S., Gupta J.K.: Theory of Machines. 14th edn. S. Chand & Co Ltd., 1071 p. (2005)
5. Uicker, J., Pennock, G., Shigley, J.: Theory of Machines and Mechanisms. 4th edn. Oxford University Press, Oxford, New York, 900 p. (2011)

6. Kosenok, B., Balyakin, V., Krylov, E.: Method of closed vector contours for teaching/learning MMS. In: García-Prada J., Castejón C. (eds.) New Trends in Educational Activity in the Field of Mechanism and Machine Theory. Mechanisms and Machine Science, vol. 64. Springer, Berlin/Heidelberg, Germany, pp. 3–10 (2019)
7. Krylov, E., Egorova, O., Gubert, A., Devyaterikov, S.: SIOMMS: evolution and development. Mech. Mach. Theory **153**, 104029 (2020)
8. Ceccarelli, M.: Innovation challenges for mechanism design. Mech. Mach. Theory **125**, 94–100 (2018)

Chapter 5
Flipped Classroom and Technology Enhanced Learning in Mechanical Engineering

A. Diez-Ibarbia, A. de-Juan, P. Garcia-Fernandez, A. Fernandez-del-Rincon, M. Iglesias, J. Sanchez-Espiga, and F. Viadero

Abstract A change in the pedagogical approach and techniques is established in two subjects of the Mechanical Engineering Degree of the University of Cantabria, in order to solve problems detected during the last decade and accentuated with the pandemic. Specifically, in first pilot, two main changes are to be performed with respect to previous years. A mixed pedagogical approach is adopted, where both the Traditional and Flipped Classroom are combined, making the class time more valuable and engaging better with the students, and a collaborative project to an "open-problem" is proposed for the students to solve in groups of three, where the majority of the competences are exercised. In this regard, the use of Technology Enhanced Learning has been seen essential in order to facilitate students' access to the knowledge and contents. In University of Cantabria context, Kaltura tool was used for this purpose, which is embedded in Moodle platform. In the second pilot, students currently have the contents of the course in "slide" format before the class sessions, therefore, the generation of audiovisual resources will consist of the complement and support of the current teaching mechanisms used. In order to aid a better student follow-up, a one-year action plan has been developed. Within this plan, it is the second month of the project and therefore, the need of the audiovisual resources has been outlined and some videos have been created.

Keywords Flipped classroom · Technology enhanced learning · Mechanical engineering

---

A. Diez-Ibarbia · A. de-Juan (✉) · P. Garcia-Fernandez · A. Fernandez-del-Rincon · M. Iglesias · J. Sanchez-Espiga · F. Viadero
University of Cantabria, Los Castros s/n, Santander, Spain
e-mail: ana.dejuan@unican.es

J. C. García Prada et al. (eds.), *Trends in Educational Activity in the Field of Mechanism and Machine Theory (2018–2022)*, Mechanisms and Machine Science 128,
https://doi.org/10.1007/978-3-031-25730-8_5

5.1 Introduction

In the last two decades, the European Higher Education Area (EHEA) has promoted a change in educational paradigm: from teaching-oriented to learning-oriented, promoting therefore different pedagogical approaches and modes for teaching and learning [1].

Until the entry into force of the EHEA, the "Conductivist" current, based on the fact that the center of the teaching–learning process is the teacher, dominated university teaching in Spain. The lecturer characteristics (such as ability, personality or cultural values themselves) are considered fundamental determinants of results or achievements in the classroom. Therefore, scientific studies based on this current focus on the identification of which qualities are necessary to be an effective teacher, such as objectivity, empathy, interpersonal sensitivity, flexibility, enthusiasm or expressiveness. Despite the fact that the Conductivism current has been followed mainly until the end of the twentieth century, it has the great shortcoming of leaving out the context variables associated with the student. Furthermore, it does not frequently stimulate and motivate students [2–4] who, in the end, do not acquire the competences, learning outcomes, skills and knowledge that the labor market requires [5].

In this sense, the current of "Cognitive" theories advances. This approach moves the center of the teaching–learning process to the student, since it considers the student as an active and essential part of the knowledge construction process [6]. In this regard, knowing how students learn and establishing a rational teaching plan, capable of stimulating the most appropriate strategies to transmit the contents or curricular subjects, become two important lines of action to face any process of improvement in the university classroom. Within the current of Cognitive theories there are several lines of research; for example, the theory of learning by construction [7], which defends that learning is carried out by inserting information into the network of knowledge and own experiences; or the theory of social learning [8], which adds to the previous one the condition that learning is inseparable from the situation in which it occurs. Furthermore, several methodologies and techniques have emerged during the last decades in this line, such us flipped classroom [5, 9], problem/project-based learning [10], inquiry-based learning [11], work-based learning [12], technology enhanced learning [13–15], etc., that, in the end, aid students to obtain the expected competences. All of them focused on achieving an active role of the students and increasing their motivation. This fact acquires special importance during the development of the learning process. It is evident that no one will learn if they do not want to learn [16]. However, the will to learn can be activated, inhibited or limited by direct actions of the teaching staff, which is why it constitutes a fundamental factor in the teaching approach choice.

These two fundamental currents should not be considered as individual options, but rather the opposite. Trying to integrate both should be the goal, since there is no single valid theory for all cases. What must be clear is that the role of the teacher should no longer be a task with a unidirectional meaning (transmission from the teacher to student) and, therefore, it does not consist solely of the oral presentation of a series of knowledge, nor the delivery of written information to students. Although these kinds of activities are necessary for the true teaching function, of course they are not enough, and it is necessary to contribute with something more.

This becomes essentially these days, just after the COVID-19 pandemic situation, which has aroused a latent issue: student's absenteeism to onsite lessons. Authors conducted a survey to all Mechanical Engineering Degree (MED) students; which goal was to learn the reasons that motivate their nonattendance to face-to-face (F2F) sessions. The results shed some light to this particular issue, highlighting one common answer shared by the majority of the students: they perceive that some instructors just read the slides in the onsite lessons, which turn their presence into useless.

In this work, two pilot experiences, which are being performed in two subjects of Mechanical Engineering Degree in the first semester of academic year 2022/2023, are presented. The aims are for students to obtain the competences which are required in their future jobs, to increase their motivation in Higher Education Curricula, and to reduce the absenteeism. These goals are planned to be achieved with the help of the University of Cantabria (UC) participation in a European Project denominated as "E-desk project", which *believes in the importance of having proficient university teachers in digital and entrepreneurial competencies to enhance European Youth's lifelong learning, improve its employability and foster the European values.* This project, under the EntreComp framework that promotes the entrepreneurial competences [17], fosters the creation of audiovisual teaching resources, as support for F2F teaching in the classroom and even as eliminating the read-the-slides problem, because the first two levels of Bloom's taxonomy revised in 2001 [18], namely, remember and understand, can be achieved previously to the onsite lessons. The envisaged result is to facilitate the understanding of the contents, which can be adapted to students' learning pace and to their needs and abilities, promoting self-learning and improving accessibility for students with specific educational support needs. It is also expected to increase the student's motivation because they feel more comfortable with digital contents in their daily activities and leisure.

5.2 Context of the Two Pilot Experiences

Firstly, the Mechanical Engineering Degree of the UC where the two subjects belong, is briefly described. It is also summarized the plan and the subjects that are advisable the students have passed before these two (Fig. 5.1). Furthermore, the competences that students will acquire once they pass the subjects.

The Report [19] of the degree establishes a classification based on its level of specialization, distinguishing, in addition, the transversal competences. This document includes the skills that students must acquire, and which are transcribed in Table 5.1.

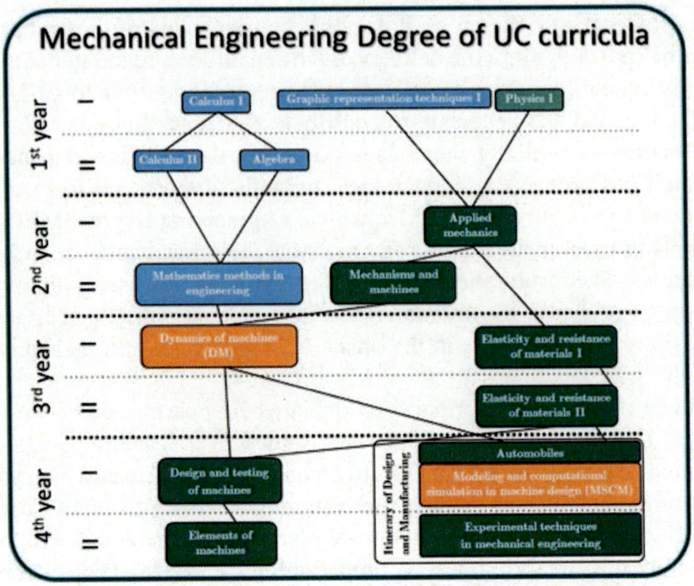

Fig. 5.1 The two subjects (orange) in the mechanical engineering degree of UC plan

Table 5.1 Competences students must acquire in the two subjects

Code	Description	Subject
ITI_GT3	Knowledge in basic and technological subjects, which enables them to learn new methods and theories, and gives them the versatility to adapt to new situations	DM, MCSM
ITI_GT4	Being able to solve problems with initiative, decision-making, creativity, critical reasoning and to communicate and transmit knowledge, skills and abilities in the field of mechanical engineering	DM, MCSM
ITI_TM1	Knowledge and skills to apply graphic engineering techniques	MCSM
ITI_TM2	Knowledge and skills for the calculation, design and testing of machines	DM, MCSM
GTRA4	Problem solving	DM, MCSM
GTRA7	Being able to communicate verbally	MCSM
GTRA13	Being able to work in a team	MCSM

To exercise the competences that students must acquire, lecturers have to plan in advance the contents, activities and assessments in which our students apply the information from our discipline to the resolution of relevant problems [9]. Taking this into account, the two pilot activities developed by the authors are presented next. Each one will be tracked in real time, in order to check the participation of the students when an activity is proposed. Two weeks after each activity, another control will be performed. By the end of the course, a survey will be distributed among the students in order to check their subjective perception about the competences and learning outcomes they have acquired, which will be based on previous experiences [20, 21].

5.3 Pilot Experience 1: Modeling and Computational Simulation in Machine Design

This subject, entitled "Modeling and computational simulation in machine design", belongs to the 7th semester of the degree and is eligible, which means that not all students will select it as part of their studies. It usually has around 25 students enrolled each academic year. It is divided into two blocks of contents: Finite Element Method (FEM) and Multi-Body Systems (MBS). The former is the only one partially changed at this point, since, first, validating the methodology is required and, second, in the case of a full subject transformation, making it smoothly is advisable.

Two main changes are to be performed with respect to previous years. Firstly, a mixed pedagogical approach is to be implemented, where both the traditional and Flipped Classroom (FC) are combined. Secondly, a collaborative project to an "open-problem" is proposed for the students to solve in groups of three, where authors have previous experience [20, 21].

In order to understand better the changes introduced to the FEM block of contents, the Old Plan (OP) of the course is compared with the new one (NP) and presented in Tables 5.2 and 5.3. In these tables, the half-of-semester-timeline is represented by the fifteen class sessions (2-h sessions) and the contents are listed by the theory and practice activities. The subject start with a session of theory where the context is established and then the rest of theories and concepts are incorporated and applied in the practical activities throughout the half of the semester. Moreover, orange is the Work in Class (WC) F2F time, blue the Individual Autonomous Work (IAW) and green the Collaborative Autonomous Work (CAW) both outside the class.

It can be seen that in the OP, as traditional approach was mainly used, the students did not have to make much IAW in understanding the concepts, since they were explained and practiced in the class time. Their IAW was mainly dedicated to solve

Table 5.2 Old plan of FEM block (Traditional) versus new plan (mixed of traditional-FC)

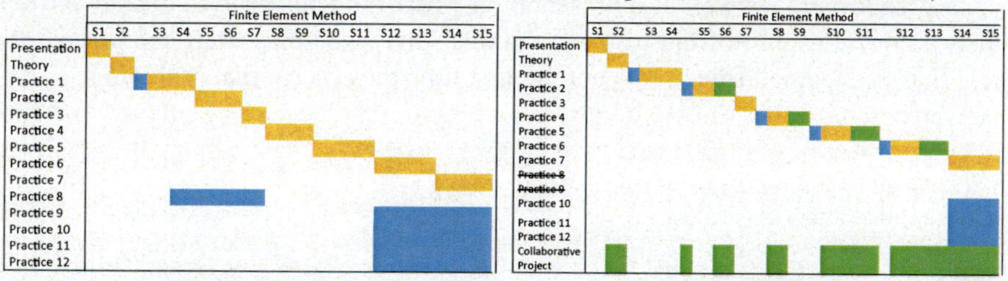

a) OP of FEM block

b) NP of FEM block

Table 5.3 Comparison between old plan (OP) and new plan (NP)

	Sesiones	
	OP	NP
WC	15	11
IAW	21	11
CAW	0	14
TOTAL	36	36

five practices and write a report of each one. In these reports, students must answer what the problem is, how they solve it and why, analyzing the results and outlining some conclusions.

On the other hand, in the NP, a mixture of traditional and FC approaches is established. Currently, students have to make IAW to prepare the concepts in advance of some practices, otherwise they will not be able to assess the results properly. In this way, some steps of the problem solving, which had to be repeated to obtain the FEM model, are not performed during class time. This saves approximately ten hours of class time, of which two are currently dedicated to outline conclusions and solve interesting questions (students come with the lesson learned before starting the class) and the remaining eight are used to guide during F2F time the student groups and aid with the Collaborative Project (technical and report issues).

To include FC pedagogical approach, the use of Technology Enhanced Learning is advisable, in order to facilitate students' access to the knowledge and contents. In this regard, in previous years, students had the contents uploaded in Moodle platform in pdf format, and the lecturer made them available sequentially with the pace of the F2F classes. To aid this FC approach, delivery mode implemented is blended, which means that some contents must be acquired outside the class. In this line, some self-explained videos where recorded with Kaltura tool, which is embedded in Moodle platform.

At the beginning of the semester (presentation session), students are informed that they have to perform, in groups of three, the design of a component, in this case a chair. As it is an open problem, they have to choose the application (spot opportunities) and technicalities (size, movement or not, material, external forces, etc.), using the tools and knowledge acquired. With this collaborative project, first, all the competences are exercised (ITI_GT3, ITI_GT4, ITI_TM1, ITI_TM2, GTRA4, GTRA7, GTRA13). Second, most of the entrepreneurial competences are also necessary to be used [17, 22]. Lastly, the report of this collaborative project has the specific format of the End-of-Degree Project, as well as a presentation where students have to defend and sell their product/design in front of a jury. In this way, when they draft and present the real End-of-Degree Project, it is not the first time they face a real problem.

5.4 Pilot Experience 2: Dynamics of Machines

This second subject, entitled "Dynamics of machines", belongs to the 5th semester of the degree and is mandatory for all students. It usually has around 50 students enrolled each academic year. In order to have the proper context of the subject, first, the summary of the contents is presented in Table 5.4.

Students currently have the contents of the course in "Slide" format, on which they take notes in the sessions. This generation of audiovisual resources will consist of the complement and support of the current teaching mechanisms used. For this, the use of the Kaltura tool is proposed, with which explanatory videos of the taught concepts will be generated and developed, as well as computer-assisted simulation or laboratory application examples, in which the student visualizes and relates analytical formulations with the real physical phenomenon.

In order to aid a better student follow-up and complement and support the current teaching mechanisms, an action plan has been scheduled, which consists of initially identifying the need for the audiovisual resource within the contents of the course. Subsequently, the authors have to design, plan, carry out and record representative application examples of the course contents. This stage consists of numerical simulation, using specific mechanical design software (multibody software such as Adams or Working model or finite element software such as Nastran/Patran), or carrying out experimental tests in the laboratory of the Mechanical Engineering group (analysis experimental modal or vibration measurement and control), which also will help the instructors to show their research work in the latest projects. Once the relevant recordings have been made, the editing and processing of the audiovisual resources

Table 5.4 Contents of the subject

Block of contents I: Rigid-body dynamics	
Unit 1	Direct and inverse dynamic problem
Unit 2	Flywheels
Unit 3	Balancing of rigid rotors
Unit 4	Gear dynamics
Block of contents II: Vibrations theory	
Unit of contents II.1: Discrete systems	
Unit 5	Free vibrations of single-degree-of-freedom (DoF) systems
Unit 6	Forced vibrations of single-DoF systems
Unit 7	Transmissibility, vibrations isolation and damping measurement in 1DoF systems
Unit 8	Vibrations in two and multi-DoF systems
Unit of Contents II.2: Unidimensional continuous systems	
Unit 9	Beam vibrations under axial, torsion and bending stress
Unit of Contents II.3: Random and control of vibrations	
Unit 10	Introduction to random and control of vibrations

is proposed, incorporating the explanatory discourse of the teachers involved in the project. Finally, the teacher responsible for each subject in which these audiovisual resources are used must enable them in a timely manner, as well as organize them within the virtual classroom (Moodle platform) of the courses involved.

5.5 Conclusions

A change in the pedagogical approach and techniques is established in two subjects of the Mechanical Engineering Degree of the University of Cantabria, in order to solve problems detected during the last decade and accentuated with the pandemic. The two pilot experiences are work in progress, so no general conclusions are available by the deadline of this call of papers. Nevertheless, some outcomes can be stated at this point; in order to fully embrace the change of mentality, both instructors and students must be willing to do it. It has been noticed up to this point by the monitoring of the visualization of virtual content, that students are used to work within the frame of the traditional approach in most subjects of the degree. So, when their studies are coming to an end and there are some subjects that are shifting the paradigm, they are reluctant to do it. There are some reasons than can motivate this situation; because usually it means more work for them and because they are simply not used to it.

Acknowledgements The authors would like to also acknowledge projects PID2020-116213RB-I00 and PID2020-116572RA-I00 funded by the Spanish Ministry of Science and Innovation and the 1st Call of Audiovisual Resources Generation of University of Cantabria.

References

1. European Higher Education Area. http://ehea.info/index.php
2. Mohr, K.A.: Understanding generation Z students to promote a contemporary learning environment. J. Emp. Teach. Excel. **1**(9) (2017)
3. Arum, R. et al.: Academically Adrift: Limited Learning on College Campuses. The University of Chicago Press (2010)
4. Bok, D.: The Struggle to Reform Our College. Princeton University Press (2017)
5. Murillo-Zamorano, L.R. et al.: How the flipped classroom affects knowledge, skills and engagement in higher education: effects on students' satisfaction. C & E, **141** (2019)
6. Ausubel, D.P.: Adquisición y retención del conocimiento: Una perspectiva cognitiva. Ed. Barcelona: Paidés (2002)
7. Piaget, J.: La equilibración de las estructuras cognitivas: Problema central del desarrollo. Ed. Siglo XXI (1978)
8. Vygotsky, L.S.: El desarrollo de los procesos superiores. Ed. B: Crítica (1979)
9. Prieto, A., et al.: La enseñanza universitaria basada en la actividad del estudiante: evidencias de su validez en docentes universitarios. Una formación centrada en la práctica, Madrid: Porlán. R., de Alba Fern (2020)
10. Liu, X., et al.: Practice of project-centric flipped classroom learning in microcomputer interfacing technology course (2017)

11. Chiang, T.H.: Analysis of learning behavior in a flipped programming classroom adopting problem solving strategies. Interact. Learn. Environ. **25**(2) (2017)
12. Nouwen, W., et al.: The role of work-based learning for student engagement in vocational education and training: an application of the self-system model of motivational development. Eur. J. Psychol. Educ. **37**(3) (2022)
13. Dunn, T.J., et al.: Technology enhanced learning in higher education; motivations, engagement and academic achievement. Comput. Educ. **137** (2019)
14. Mukhemar, R., et al.: Effect of implementing technology-enhanced learning (TEL) on students' motivation—a literature review (2022)
15. Zhang, Y., et al.: The relationship between technology leadership and technology-enhanced teaching and learning engagement in higher education. In: 7th International Conference on Distance Education and Learning, China, vol. 1 (2022)
16. Garcia Hoz, V.: La educación personalizada en la Universidad. Ed Rialp (1996)
17. European Commission, Employment, Social Affairs & Inclusion Directorate-General for Employment, EntreComp: the European Entrepreneurship Competence Framework (2019)
18. Anderson, L.W., et al.: A Taxonomy for Learning, Teaching, and Assessing: A Revision of Bloom's Taxonomy of Educational Objectives. A & B (2001)
19. Memoria para la Solicitud de Verificación de Títulos Oficiales de GIM (2008)
20. del Rincón, A.F., et al.: Application of an inter-university competition on the enhancement of Engineering Degrees (2013)
21. de-Juan, A., et al.: Enhancement of mechanical engineering degree through student design competition as added value. Considerations and viability. J. Eng. Des. **27**(8) (2016)
22. e-DESK (Digital and entrepreneurial skills for European teachers in the COVID-19 World). Erasmus+ (2022). https://edeskeurope.eu/

Chapter 6
Flipped Learning Applied to Machine Design

Juan Carlos Jauregui-Correa

Abstract This paper presents the Flipped Classroom Methodology applied to teaching Machine Design. The document answers a question regarding teaching Machine Design in virtual media. The Flipped Classroom Methodology combines new technological tools with disruptive teaching concepts, plus the student's ability to search for information within the net. The teacher's role became selecting the most appropriate reading material, the relevant examples and exercises for working at home, and the definition of classroom activities for shearing and socializing the knowledge. In this way, the students should exchange their learning experiences, and their doubts become the center of discussion. The teacher is no longer the knowledge supplier but instead leads the group to find the answers. The flipped classroom methodology extends the classroom environment beyond the traditional classroom limits. The methodology links, through the new internet technologies, the available sources of information, the student's timing, and the possibility of the shearing learning experience. This paper presents the experience of using the flipped classroom methodology for teaching Machine Design.

Keywords Flipped classroom · Active learning · Distance education

6.1 Introduction

This section describes the Flipped Classroom Methodology from a general pedagogic perspective and emphasizes its application to teaching Machine Design during the pandemic. The methodology was the most appropriate teaching recourse to overcome the difficulties imposed by the need for isolation and home office. The decision to use this method was a response to the imminent need to deliver the courses from home. Several alternatives were sought to face the academic challenge without an apparent reference, local background, or close people with whom to consult. The

J. C. Jauregui-Correa (✉)
Autonomous University of Queretaro, 76010 Santiago de Querétaro, Querétaro, Mexico
e-mail: jc.jauregui@uaq.mx

J. C. García Prada et al. (eds.), *Trends in Educational Activity in the Field of Mechanism and Machine Theory (2018–2022)*, Mechanisms and Machine Science 128,
https://doi.org/10.1007/978-3-031-25730-8_6

selection of the flipped classroom methodology came from direct observation of online courses. The first option was the asynchronous methodology, which limits any direct interaction between the teacher and the students or even among students. Asynchronous models work very well when students have a solid background and know how to interpret the information and perform the exercises, and the assessment methods are well-defined. It also allows group activities and the creation of discussion forums; the limitations were the way to complement collaborative work and the solution of specific doubts.

The other teaching option was changing the classroom ambiance for a series of presentations and giving the lecture online. From observing other teachers and consulting with students, it was clear that this method was ineffective. Students can be distracted by other online applications and copy the solution from the online tools. It was also observed that the students were not immersed in the class; they only waited to receive the material and then tried to understand it and use it to solve the assigned work and exams. The conclusion was that this method proved to be the least effective for the students to assimilate new knowledge and that they couldn't fulfill the course's objectives. On the other hand, the need to teach online classes opened up many opportunities and made it possible to change a fundamental paradigm in engineering education "to promote skills so that students become self-taught."

The search for alternatives found that the Flipped Classroom Method could overcome some limitations of teaching online courses. This concept allowed structuring the daily work with a double objective: the students understand the theoretical background of Machine Design and strengthen the habit of self-study without ceasing to have a direct teacher-student interaction.

6.1.1 Connectivism and the Flipped Classroom Method

It is essential to define some concepts. The Flipped Classroom Method is based on a pedagogical principle defined as connectivism. Connectivism starts from generating connections between the different knowledge actors and their sources. To do this, it seeks to identify the information networks and their connection with the context in which they work and build knowledge through socialization and collaborative work among the participants. Its elements can be summarized in reference [1]:

1. Knowledge lies in the diversity of opinions
2. Knowledge is built by connecting information nodes
3. Knowledge is intrinsic to the environment and does not only depend on humans (it is in the sources of information)
4. It is more important to know how to increase knowledge than what is known (we do not need the data, but to find it on the network at the right time)
5. Continuous knowledge is nurtured by maintaining and nurturing connections
6. It is essential to have the ability to see the connections between the fields of knowledge and ideas

7. The connectivism process intends to have accurate and up-to-date learning material
8. The selection of the learning material and its meaning must be seen from a different perspective.

The teacher's role is to create a learning environment in which the student becomes the owner of his knowledge. The tools are blogs, wikis, and networks in which the learning process takes place. In a community, meanings and texts (videos) are constructed whose content is exchanged and negotiated, achieving the collective creation of knowledge. This concept was not contemplated in traditional engineering education and is a new challenge for teaching in a virtual environment [1].

Ortega [2] described Siemens' work on behaviorism, cognitivism, and constructivism, and determined their limitations, since they were not developed when technology had not had an impact on learning at the current level; databases and information available on the network are, for the above parameters, infinite. Not so with knowledge, since today there is access to exorbitant amounts of data, but not necessarily knowledge. These theories were developed when knowledge grew more slowly. Knowledge increases dramatically, and learning no longer follows the same process as before; now, it occurs in different ways and scenarios. Thus, learning must be seen as a process based on a close relationship between the object of study and the tasks the student must perform in an open and global environment. New learning processes require systems to be available to information and capable of classifying and ordering interactions within the learning environment. Plans must be flexible to adapt to change and take advantage of experience in an evolutionary way. The new knowledge is not passive; it is active with all the actors and is created when a community is capable of transforming the amount of information into something useful for the group. The process is no longer linear; it is no longer based on a message sender, a means of transmission, and a receiver. Currently, each individual in a group is part of an active network in which each element of the group has access to a sea of information, which must be shared so that a new piece of knowledge can be organized, analyzed, and built. The new process depends on the teacher's role of ensuring that the interactions occur, that the organization of the information follows the individual development of each participant in the group, and promotes the mechanisms for sharing the results of the collective analysis. This concept can be applied more easily thanks to the technological resources of the modern world because it facilitates the interconnectivity of all the group members [2].

The elements of the flipped classroom [3, 4];

- Build a flexible environment in which the student accesses the material and self-manages it
- Student autonomy implies that the student is: autonomous, participatory, collaborative, dynamic, and interactive
- The teacher ceases to be an exhibitor and becomes a guide. Create content, adjust the content according to the needs of the group, establish the levels of academic quality and allow diversity in the progress of the group.

The inverted classroom requires the support of the family since the activities are also done at home.

6.1.2 The Flipped Classroom Method

The model puts the student at the center of the process and assumes responsibility for learning, so the teacher takes another role and prepares the learning material through digital media. The teacher becomes responsible for designing the materials and activities the student must develop at home. The didactic methodologies must be chosen to create coherence between the contents and the activities.

The student is responsible for learning.

- The teacher is responsible for content design and consistency between content and activities
- The working tool is the digital communication media.

The pedagogical approach is that instead of having group instruction (classroom), instruction is individual (homework). In the (virtual) group space, students build the experiences defined in the activities in a collaborative environment. The role of the teacher is that of a dynamic accompaniment, generates and motivates creative conditions that point to innovation, and gets involved with the students in constructing new concepts and their application.

Basis of the method

- Flexible environment: Accompany the autonomous learning process, organize the space in the virtual platform, prepare the materials, and define the activities according to the needs of the students.
- Assessment: Create adequate spaces for interaction, and reflect on their learning. Observe, follow up with students and adjust when necessary. Offer different ways to learn and demonstrate experience of the subject.
- Culture of learning: The responsibility for instruction is centered on the student, who assumes responsibility for learning. Through activities, this knowledge is deepened, reinforced, and recognized in the classroom.
- Directed content: The material is generated in a good way; it adapts to the needs of the students, creates interest, and allows the student to identify their strengths and weaknesses to assimilate each unit. The material has an educational purpose, so the teacher has to curate (select and design) the information and adopt active learning activities that promote independence, autonomy, and responsibility in the student.
- Teacher's work: Guide, facilitator, curator of materials, stays close, accompanies, and provides feedback to students to promote an environment of critical self-reflection of the knowledge learned. It is necessary to generate reflection among peers so that the teacher can improve their guiding process.

Advantages of the flipped classroom:

- Improves the work environment in the classroom
- Increase educational attention to each student
- Transform the living room into an active work area
- Promotes creativity and critical thinking
- Promotes collaborative work the content is accessible
- Parents can know what happens in class.

The final objective of the Flipped Classroom Method is to select and analyze the information that the student will have to study at home and to design the activities so that he develops deductive, analytical, and problem-solving abilities with the available information and reinforces the basic knowledge of each subject at hand through learning assessment elements [5–8].

6.2 Application to Machine Design Class

This section describes applying the Flipped Classroom Method to a Machine Design course. The first part consisted of selecting the fundamental material for the semester. Figure 6.1 illustrates the thought cycle with which engineers' skills and abilities are formed. There are three groups: Deduction, Analogy, and Reinforcement.

Knowledge and skills are built from previous experiences that, with deduction skills, generate new knowledge. This knowledge is based on analogies with similar phenomena or problems and is reaffirmed by repetition and mechanization (reinforcement).

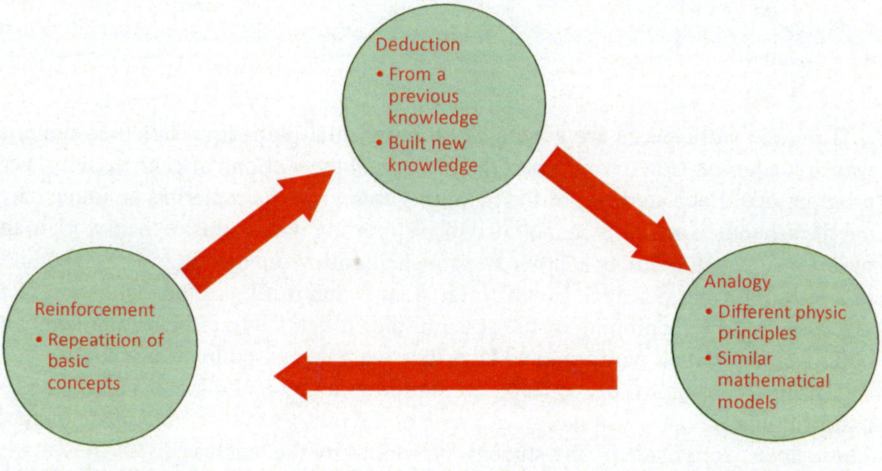

Fig. 6.1 Thinking cycle for creating engineering knowledge

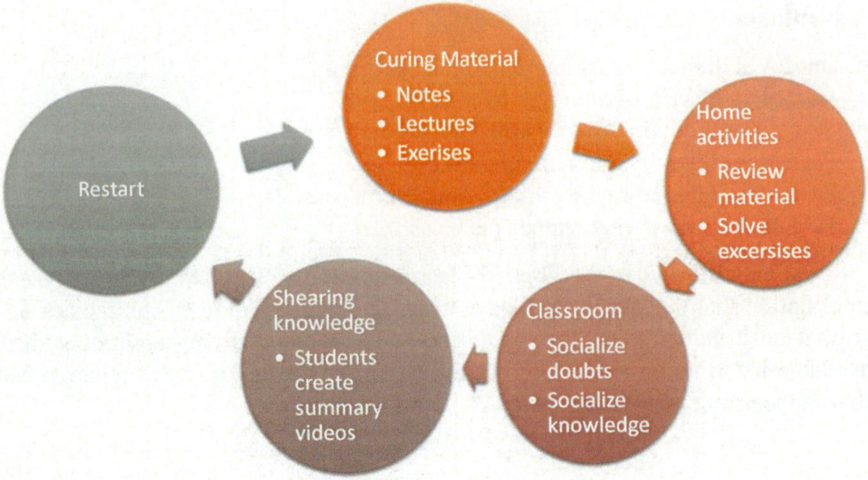

Fig. 6.2 Work flow using the flipped classroom method

The knowledge cycle within the Flipped Classroom Method has the elements shown in Fig. 6.2.

The work process is illustrated in Fig. 6.3. The cycle begins when the teacher assigns the activity and defines the delivery of each student's work. Examples of these activities are textbook readings or the formulation of theoretical-practical questions that the student must present in a formal format. Formal presentations ensure that the student develops the assertive communication skills he will need throughout his professional life.

The cycle begins with the selection and curation of the material:

- Notes
- Readings
- Exercises.

Then the workspaces are generated on the virtual platform. The cured material was uploaded on Classroom®, and the individual evaluations of each activity were programmed. Each student has the flexibility to review the materials at home, carry out the scheduled activities and upload the supporting documents of their work to the platform. Once the activity at home was finished, follow-up meetings were scheduled in a virtual format (Meet®). In which the doubts and results of the evaluation were discussed. At the beginning of the session, the students wrote their comments and doubts on the virtual platform, and later they were discussed in groups.

During the sessions, the concepts were reinforced by making notes on the digital platform. Figure 6.4 illustrates an example of the messages that were made in class. These notes were made by the students as well as by the teacher.

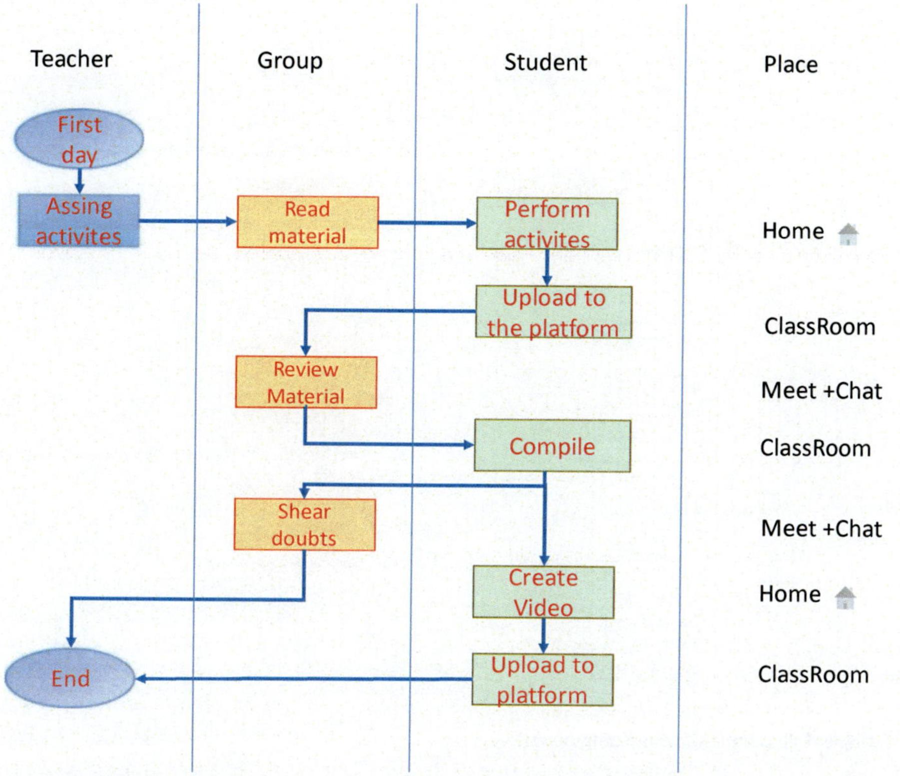

Fig. 6.3 Work flow strategy

6.3 Conclusions

It was possible to meet the course's objectives in adverse circumstances by applying the Flipped Classroom Method. The students built for themselves the theoretical foundations of Machine Design. They could evaluate their capacity for self-taught training and collectively share their doubts and progress in handling the topics. The most important contribution of this method was that the students built their deductive capacity and applied the analogies from their previous knowledge.

The method also allowed students to reinforce each concept studied through revision exercises. The collectivization of knowledge was achieved through two tools: sharing doubts and learning in writing in virtual meetings (face-to-face but at a distance) and feedback between them from editing dissemination videos (made by the students).

Since applying the Flipped-Class method, students have shown better grades compared with previous courses. Before using the technique, the average grade was 75%; last semester, students improved their performance up to 87%. There is a

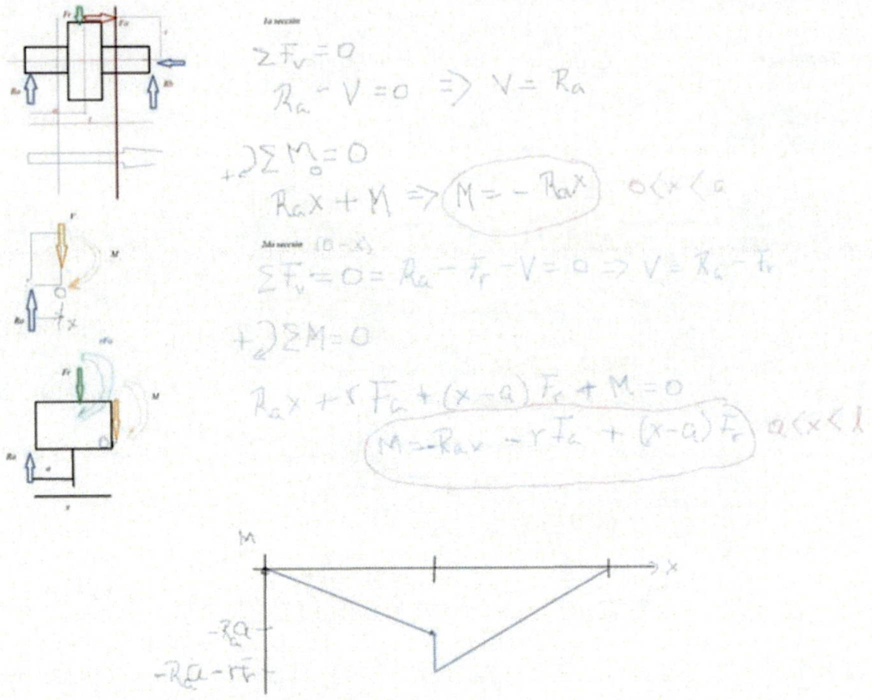

Fig. 6.4 Example of socializing doubts

need to evaluate the method from the student's perspective. This open issue will be implemented at the end of this semester.

The ability to adapt and see how the students managed to be self-taught was the most outstanding achievement of this method. The significant limitations were given by confinement: on the one hand, not being able to carry out practical activities interspersed with theoretical topics, and on the other hand, the lack of social contact between students that have effects on the formation of social ties and integration into groups that characterizes generations of engineers.

References

1. Gutiérrez, L.: Conectivismo como teoría de aprendizaje, conceptos, ideas, y posibles limitaciones. Revista Educación y Tecnología (1) (2012)
2. Ortega, R.: El modelo centrado en el aprendizaje y su implicación en la formación de docentes. Revista Iberoamericana para la Investigación y el Desarrollo Educativo, vol (10) (2013)
3. Bergmann, J., Sams, A.: Dale vuelta a tu clase. Ediciones SM (2012)
4. Cho, H., Zhao, K., Lee, C., Runshe, D., Krousgrill, C.: Active learning through flipped classroom in mechanical engineering: improving students' perception of learning and performance. Int. J. STEM Educ. (2021). https://doi.org/10.1186/s40594-021-00302-2

5. Gatica-Lara, F., Uribarren-Berrueta, T.: "¿Cómo elaborar una rúbrica?", Investigación en Educación Médica (2012)
6. Prevalla, B., Uzunboylu, H.: Flipped learning in engineering education. TEM J. **8**(2), 656–661 (2019). https://doi.org/10.18421/TEM82-46
7. Vaerewyck, A.: Flip the Classroom Alum develops popular new teaching style. University of Colorado-Denver (2013)
8. Pazmiño, I.: Planificación de la clase invertida. Editeka Ediciones (2014)

Chapter 7
Dimensional Synthesis Approach of A Compliant Clutch Mechanism for a Formula Student Car (I)

Gerardo Peláez, Gustavo Peláez, Higinio Rubio, and Alejandro Bustos

Abstract Compliant or pseudo-rigid mechanisms that can provide optimal motions, carry out particular tasks, and cope with mechanical vibrations represent a challenge for its dimensional synthesis and mechanical design. These mechanisms gain mobility owing to the relative displacement of a joint or flexible-member deflection. Thus, they perform movement, force, and energy transfer during deflection. This study builds on the near-optimal dimensional synthesis of a flexible frame mechanism, by discussing the effects of a pivot point at toggle positions. Certain assumptions are made. Specifically, the customary closed-loop algebraic equation constraints are relaxed by inequality constraints.

Keywords Compliant mechanism · Near-optimal · Dimensional synthesis

7.1 Introduction

7.1.1 Literature: Brief Review

In the mechanisms design pursuit, we are continually faced with trade offs. One of such trade offs, which is evident, is the relationship between mechanism requirements and type plus the number of its links and joints. In most cases, to achieve high requirements, the number of links and joints must be increased. Throughout this study, we attempt to give up unimportant quantities to improve important performance measures. This is the case for the near-optimal dimensional synthesis of

G. Peláez (✉) · G. Peláez
University of Vigo, Vigo 36310, Spain
e-mail: gpelaez@uvigo.es

H. Rubio
University Carlos III de Madrid, Madrid, Spain

A. Bustos
UNED, Madrid, Spain

J. C. García Prada et al. (eds.), *Trends in Educational Activity in the Field of Mechanism and Machine Theory (2018–2022)*, Mechanisms and Machine Science 128,
https://doi.org/10.1007/978-3-031-25730-8_7

a four-bar pseudo-rigid mechanism with a flexible frame and we attempt to obtain several performances from very little.

The dimensional synthesis of the mechanisms that, in addition to create special routes plus positions, and also provide a specific torque gain as output, all without branches, is called rectification solution [1]. The aforementioned study builds on six-bar mechanisms. In a trial of mechanisms with a minor number of links, Plecnik and McCarthy used a random search to find non-branching slider-crank function generators [2]. Combining these behaviours with other functionalities is an active research area [3–5]. In particular, pseudo-rigid mechanisms capable of undergoing mechanical vibration at some joints, maintaining unchanged the performance to accomplish desired tasks or trajectories, offer multifunctional benefits over traditional mechanisms [6]. Pseudo-rigid or compliant mechanisms have at least one flexible link that increases their mobility owing to the relative motion of a joint that toggles [7]. A wide array of studies have considered the structure and function of the compliant mechanisms. The basic nomenclature and classification of its components are established in [8]. Howell introduces in [9] a method to aid in the design of a class of compliant mechanisms wherein the flexible sections (flexural pivots) are small in length compared to the relatively rigid sections. A formal structural optimisation technique, called the homogenisation method, to design flexible structures, that is, compliant mechanisms, was proposed by [10]. With the goal of technically realisable compliant joints, and their integration into elastically movable structures for motion tasks in positioning and manipulating engineering, new ideas were introduced in [11].

Anyway, the necessity to synthesise planar linkage mechanisms with intended functions in a broad range of industrial sectors remains unaltered. Therefore, there has been extensive research related to the mechanism synthesis and analysis. The investigations mainly focused on the type, number and dimensional synthesis [12, 13]. When more than five precision points are specified and the mechanism does not have as many design parameters, the more appealing option is to proceed with optimum synthesis numerical methods that obtain approximate solutions [14]. To this end, numerical gradient-based methodologies are promising owing to their computational efficiency [15, 16]. In this study, the General Algebraic modelling Systems (GAMS) software was selected. GAMS is particularly recommended, as we deal with a high number of non-linear inequality constraints to capture the increase in mechanism mobility. Primary work in this field was conducted by Fox and Willmert [17].

Regarding the inertial forces in a slider crank mechanism that undergoes vibration and wear and decreases its life span, many studies have been carried out on the optimisation of the dynamic behaviour of the mechanism. However, few studies have focused on the estimation and minimisation of joint forces, an exception can be found in [18]. Mobile machines composed of a planar mechanism with a single degree of freedom were developed and assembled for designing and prototyping to inspire junior and senior high school students by [19]. The present work addresses both mechanism optimisation and assembly prototyping of the near-optimal solution found for the four-bar pseudo-rigid linkage. For a more in-depth review of past and present mechanisms, readers are referred to [20, 21].

7.1.2 Design Philosophy

This work presents an optimal dimensional synthesis procedure for a four-bar pseudo-rigid linkage that generates the clutch motion of a Formula Student Car (FSC) steer by wire using a servomotor at the crank. The servomotor was operated by the steering wheel using cams. This makes driving the FSC more ergonomic and friendly to the driver. Therefore, this man-machine interface improves the vehicle performance. Figure 7.1a shows the basic four-bar linkage parallelogram to be optimised, Fig. 7.1b depicts the optimised assembly prototype in the FSC. Dimensional synthesis is based on the following principles:

- The performance specifications should be achievable in the actual implementation of the four-bar flexible frame linkage
- The neighbourhood around 'optimal' solutions should be examined for 'near-optimal' solutions that provide significant performance improvements in primary constraints measures
- The performance specifications should be considered somewhat flexible
- The four-bar flexible frame linkage should have robustness to modelling errors.

Use of Achievable Performance Specifications The four-bar linkage is considered a function generator, where the driving and driven link angular displacements are related by a prescribed correspondence. In this context, owing to the finite number of geometrical parameters of the four-bar linkage, the correspondence given by the mechanism will coincide with the prescribed only at a finite number of points, called precision points, and might deviate somewhat from the correspondence between these points. If this is the case, then six specified precision positions are selected, keeping the output torque for each position above a threshold level higher than the resistant torque of the clutch operation. The single-objective optimisation consists of maximising the rocker arm angular velocity ω_4 that is the clutch velocity. This is in conflict with the torque-gain constraint [20].

As stated above, in the pursuit of mechanism synthesis, we are faced with tradeoffs. One such tradeoff, which should be obvious from this point, is the relationship between the rocker arm angular velocity and output torque T_4. In most cases, high levels of angular velocity ω_4 cannot be obtained with a four-bar linkage, which produces a high output torque. To achieve high angular velocity ω_4, the torque gain must be decreased. In general, improving the velocity performance requires sacrificing the output torque performance. That is, we cannot obtain something from nothing. Nonetheless, throughout this study, we attempt to obtain something from very little.

The Performance Specifications Should Be Considered Somewhat Flexible Regarding Fig. 7.1b, given that the revolute joint O_4 is fixed to the Formula Student Car twin-cylinder engine ensemble, which undergoes mechanical vibrations, not to the vehicle frame, deviations from its theoretical position are expected. Compared to four-cylinder engines, twin-cylinder engines have the characteristics of small inertia and a wide range of excitation frequencies. The structural vibration and noise gen-

Fig. 7.1 The multi-rigid-body mechanism subset plus the servo-motor to steer by wire the clutch. Conditions: **a** the mechanism without optimisation and **b** optimised four-bar linkage structural layout in the Formula Student Car

erated by the twin-cylinder engine at high-speed ratios were much higher than those generated by the tetra-cylinder engine. Thus, a dimensional tolerance level must be established to deal with the actual position of the revolute joint that toggles. Therefore, instead of placing zero constraints on the closed-loop algebraic equations, it is more realistic to relax the closed-loop algebraic equations corresponding to the proposed precision points, to a low level, rather than requiring the equations to be exactly zero. To achieve this theoretical possibility of being worth exactly zero, other performance criteria must be sacrificed. Therefore, the inequality constraints were chosen.

The Four Bar Linkage Should Have Robustness to Modelling Errors The clutch system is steered by wire; therefore a servo motor acts on the crank, as shown in the overall Fig. 7.1. The Fig. 7.1a shows the initial clutch mechanism without optimisation of a four-bar parallelogram Fig. 7.1b the actual optimised four-bar linkage. The servomotor provides precision control over the crank angle φ_2; however, the input torque T_2 that is supplied is somewhat limited in this case to 10.78 $[N \cdot m]$. Therefore, in Fig. 7.2, a high torque gain is desirable to ensure that the output torque becomes greater than the clutch resistant torque by approximately 11.4 $[N \cdot m]$.

A subset of constraints was extracted from the actual mechanical design corresponding to the modelled four-bar linkage. To anticipate this concept at revolute joint B and avoid collisions between the corresponding joint bodies, a minimum value for the angle μ must be constrained, as shown in Figs. 7.3 and 7.4. If this is the case, the angle μ must be greater than the minimum value extracted from the actual mechanical design, corresponding to the angle between the connecting rod and rocker arm at joint B.

Finally, the primary assumption made in dealing with the dimensional synthesis of the four-bar linkage was that the crank, connecting rod, and rocker arm did not deform, while the mechanism frame behaved flexibly. The multirigid-body approximation suits a wide array of applications [22–25]. This is simple, allowing for faster geometric optimisation. In addition, the material and geometry of each body are specified so that the body does not deform appreciably. However the frame of the mechanism to be designed becomes a flexible link, which increases the mechanism mobility owing to the relative motion of the revolute joint O_4 that toggles, because it is joined to the engine ensemble that undergoes mechanical vibration.

7.2 Optimal Design Formulation

The kinematics and dynamics of the four-bar linkages are directly related to physical parameters. Consequently, variations in the length of the links lead to changes in the angular velocity of the beam and the output torque. This simple dependency leads to three advantages in geometric design optimisation. These three advantages lie in the ability of the four-bar linkage to satisfy the following:

1. The output torque above a threshold value.
2. The required output velocity for a given range of positions.
3. Auxiliary requirements such as smooth motion of the connecting rod for the allowable positions array.

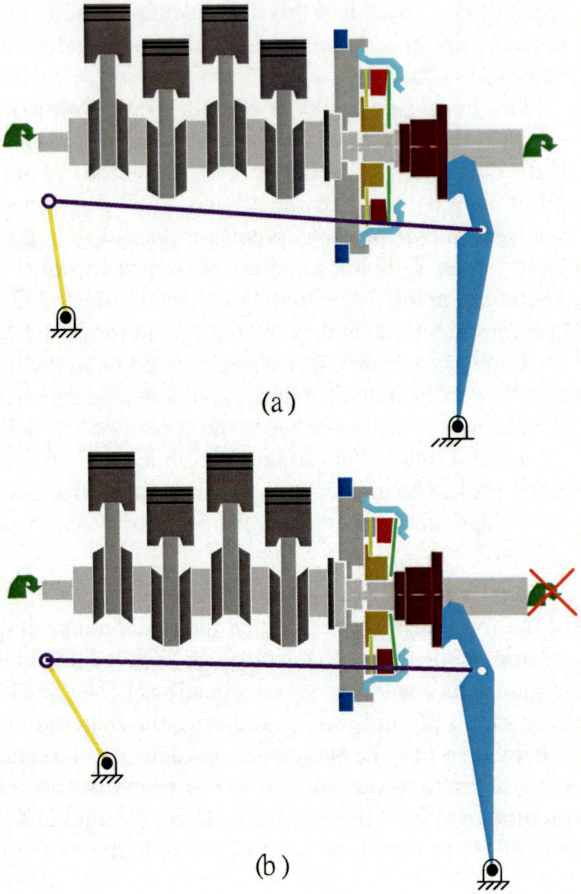

Fig. 7.2 **a** Engaged-clutch position. **b** Release position

7.2.1 *The Torque Gain Constraint*

If a four-bar linkage is designed with a given torque gain, the relationship between the applied input torque and torque generated at the output rocker arm must be specified. Once the angular velocity ratio between the crank and rocker arm is estimated, the torque gain can be easily approximated by assuming negligible energy dissipation at the joints. Several methods can be used to estimate velocity ratio ω_4/ω_2. Figure 7.5 depicts one such method, which considers a couple of effective links, the first from the pivot point O_2 plus orthogonal to the connecting rod ③ reaching the virtual point A', and the second from pivot point O_4 to point B', both parallel. Figure 7.3 also shows the principle of transmissibility:

$$v_{A'} = v_{B'} \tag{7.1}$$

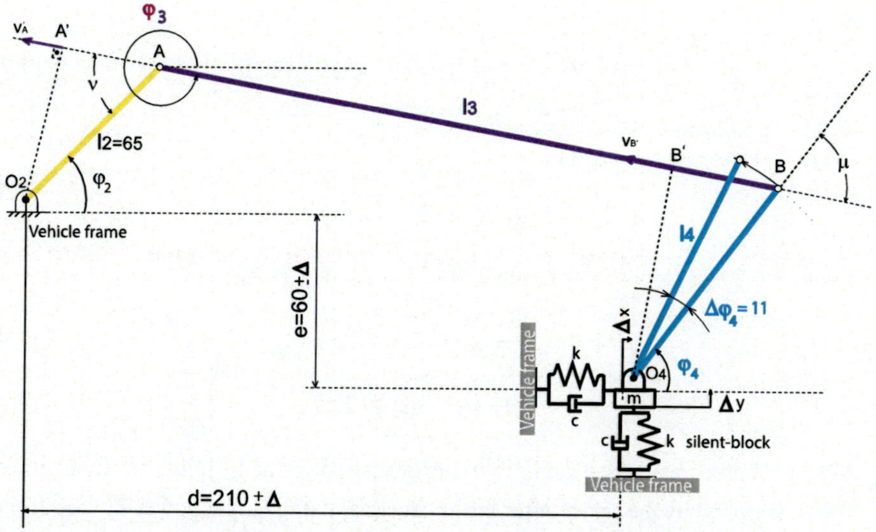

Fig. 7.3 The mechanism plus the effective links. The link lengths and the angles to be estimated are $\varphi_3, \varphi_4, \ell_3, \ell_4$. Engine amplitude of vibration Δ. Conditions: no mismatch between the axial parameters

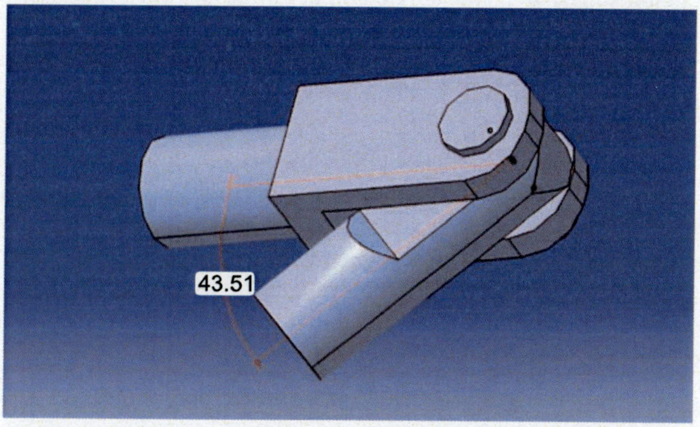

Fig. 7.4 Mechanical constraint at the revolute joint B to avoid the collisions between the joint bodies. Conditions: $\mu \geqslant 43.513°$

thus,

$$\omega_2 \cdot O_2A \cdot \sin(\nu) = \omega_4 \cdot O_4B \cdot \sin(\mu) \tag{7.2}$$

rearranging terms

$$\frac{\omega_4}{\omega_2} = \frac{O_2A}{O_4B} \cdot \frac{\sin(\nu)}{\sin(\mu)} \tag{7.3}$$

where the ν and μ values as a function of φ_2, φ_3, and φ_4 are

$$\nu = \varphi_2 - (\varphi_3 - 2\pi) \tag{7.4}$$

$$\mu = \varphi_4 - (\varphi_3 - 2\pi) \tag{7.5}$$

As stated before, assuming negligible energy dissipation at the joints, the input power $T_2 \cdot \omega_2$ supplied by the servomotor is almost equal to the output power $T_4 \cdot \omega_4$. If this is the case, then

$$T_4 = T_2 \cdot \frac{O_4B}{O_2A} \cdot \frac{\sin(\mu)}{\sin(\nu)} \tag{7.6}$$

The relationship between the applied input torque and the torque generated at the rocker arm can be easily determined by considering the output velocity generated instantaneously by a pair of gears connecting the input and output links, which is known as Kennedy's theorem, as depicted in Fig. 7.5. For a more in-depth explanation of this second method, readers can refer to [20]. The input torque available at the crank is only 10.78 $[N \cdot m]$, supplied by the servomotor. Thus, the constraint consists of keeping the output torque above a threshold value of approximately 13 $[N \cdot m]$ to counteract the resistant torque of the clutch whose estimated mean value is approximately 11.4 $[N \cdot m]$. Consequently, a desirable primary parameter combination consists of keeping the rocker arm length ℓ_4 greater than the crank length ℓ_2, regardless of the evolution of angles μ and ν, which should be controlled by the desirable set of precision points.

7.2.2 *The Auxiliary Requirements or Constraints Such as Smooth Motion of the Connecting Rod for the Desirable Positions Array*

First, as stated in the preceding section regarding the angle μ, as shown in Fig. 7.4, there is a mechanical constraint at revolute joint B. To avoid collision between the corresponding joint bodies, the angle μ should be greater than 43.513°.

Nonetheless, it is assumed as the main constraint that the four-bar linkage to be designed will accommodate the set of six precision points into its allowable

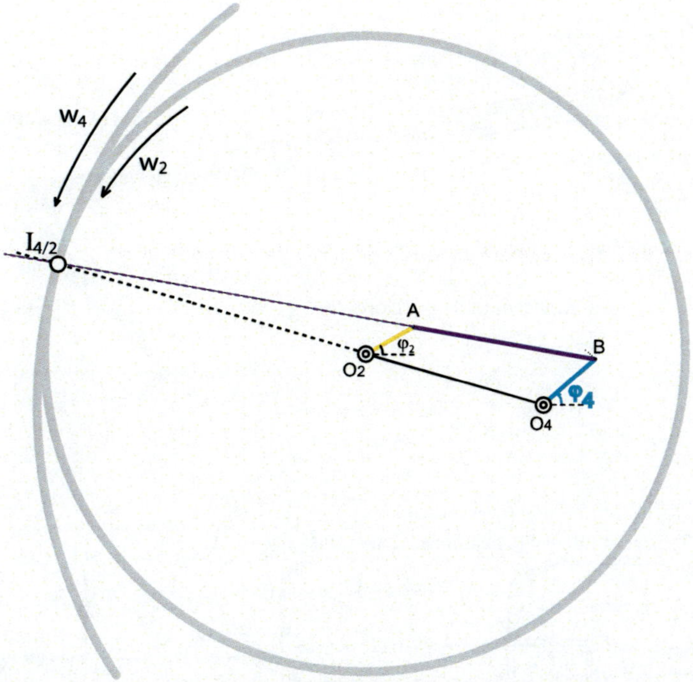

Fig. 7.5 The angular velocities of the input and output links are instantaneously equivalent to gears in contact at the instantaneous centre of rotation $I_{4/2}$

workspace. Accordingly, the customary closed-loop algebraic equations for the four-bar linkage can be written as

$$\ell_2 \cdot \cos(\varphi_2) + \ell_3 \cdot \cos(\varphi_3) - \ell_4 \cdot \cos(\varphi_4) - d = 0 \tag{7.7}$$

$$e + \ell_2 \cdot \sin(\varphi_2) - \ell_3 \cdot \sin(\varphi_3) - \ell_4 \cdot \sin(\varphi_4) = 0 \tag{7.8}$$

For each of the six allowable positions depicted in Fig. 7.6, the angle φ_4 can vary by only 11.4° over the workspace defined by the operating ranges of φ_2 and φ_3. For example, for a given increment of the crank angle ($\varphi_2 \leqslant 15\,°$), the highest variation in the angle φ_4 should be below 11.4° because the clutch cannot move more than 11.4 ° to change the plate meshing shown in Fig. 7.2. Thus, an increase of 10.5° (below 11.4 °) is preset for φ_4. Note that angle φ_2 is controlled by a servomotor that starts at 33 °. If this is the case, the angle φ_4 starts slightly above 44° for the first precision point in order to keep the angles μ and ν under well-suited values. In addition, to give equal crank angles, spacing requires:

$$\varphi_2\,[i] = \varphi_2\,[1] + (i-1)\frac{\varphi_2\,[m] - \varphi_2\,[1]}{m-1} \quad i = 1, \cdots, m \tag{7.9}$$

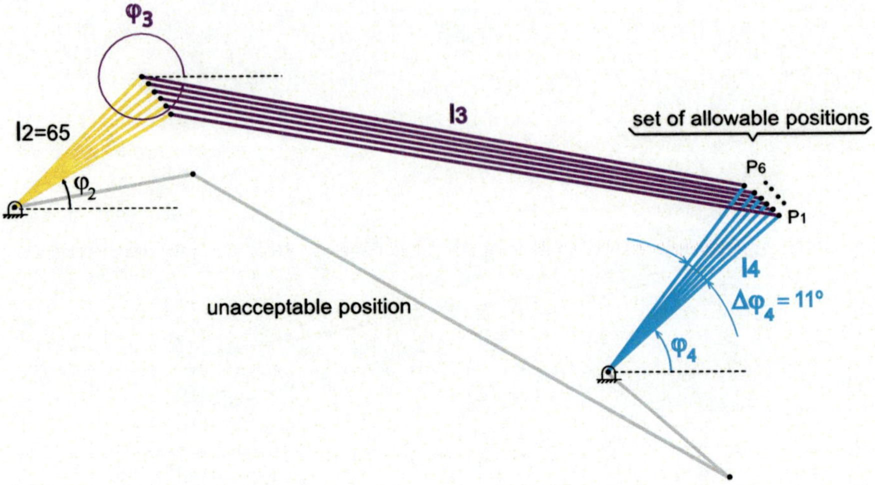

Fig. 7.6 The set of allowable positions. Conditions: $\Delta\varphi_4$=11°

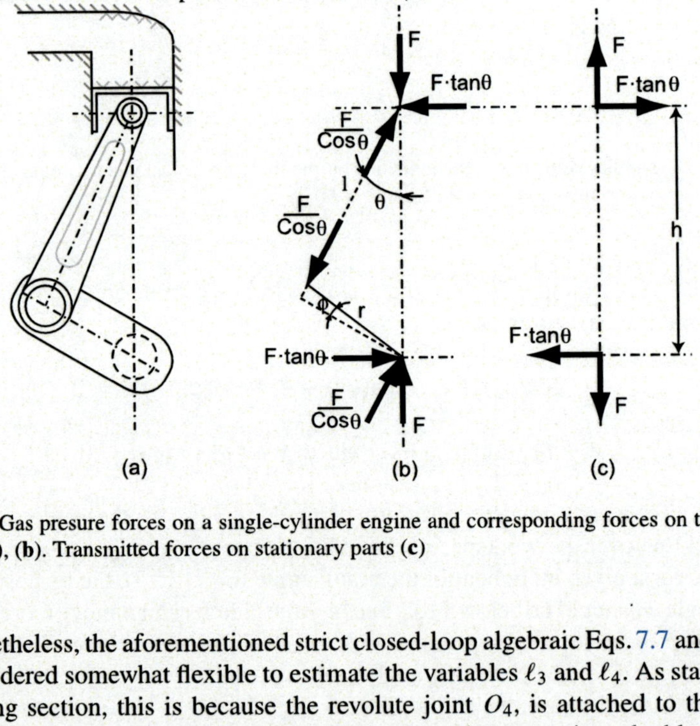

Fig. 7.7 Gas presure forces on a single-cylinder engine and corresponding forces on the moving bodies (**a**), (**b**). Transmitted forces on stationary parts (**c**)

Nonetheless, the aforementioned strict closed-loop algebraic Eqs. 7.7 and 7.8 will be considered somewhat flexible to estimate the variables ℓ_3 and ℓ_4. As stated in the preceding section, this is because the revolute joint O_4, is attached to the engine assembly and not to the vehicle frame. Thus, O_4 undergoes quite valuable displacement owing to the inertial forces that appear in the vertical and lateral directions of the engine, as shows the Fig. 7.7.

7.2.2.1 Dynamic Analysis: Inertial Forces Estimation

Fig. 7.7 represents a two-mass model of a connecting rod-crank mechanism, corresponding to a single cylinder of a twin-cylinder engine Formula Student Car. Let it be

- y_p : downward displacement of the piston from top.
- m_p : piston mass.
- ω : crank angle from top dead center.
- r : crank radius.
- ℓ : length of the connecting rod.
- λ : r/l ratio.
- M_s : Crankshaft torque.
- M_f : Reaction torque in the engine frame.
- y_c : vertical displacement of the crank.
- m_c : crank bolt mass.
- x_c : horizontal displacement of the crank.

Regarding the Fig. 7.7 the crankshaft torque can be easily estimated as

$$M_s = F \cdot r \cdot \frac{\cos(\phi)}{\cos(\theta)} \tag{7.10}$$

or equivalently about the fixed frame, the stationary part in the opposite direction M_f equals M_s. Thus, the gas pressure in the cylinder does not cause any resultant forces on the engine frame but produces only the torque about the crankshaft axis. It is also logical that at low engine speeds, the most important forces are those due to the gas. In a cruise regime, the forces of inertia reach important values regarding those due to gas pressure, and at a high rate, the inertia forces are the most important in dealing with the mechanical vibration. Thus, the first step in the discussion on the subject consists of the derivation of expressions for the vertical and lateral inertial forces for a single connecting -rod crank mechanism [26]. The goal of this section is to yield an analytical expression for inertial forces. First, the position of the piston in terms of the angle ωt is estimated.

$$y_p = \ell + r - \ell \cdot \cos(\theta) - r \cdot \cos(\omega t) \tag{7.11}$$

The auxiliary angle θ can be expressed in terms of ωt by noting that

$$\sin(\theta) = \frac{r}{l} \cdot \sin(\omega t) \tag{7.12}$$

Thus the piston mass m_p vertical displacement y_p in terms of the crank angle ωt becomes:

$$y_p = \left(r + \frac{r^2}{4\ell}\right) - r\left[\cos(\omega t) + \frac{r}{4\ell}\cos(2\omega t)\right] \tag{7.13}$$

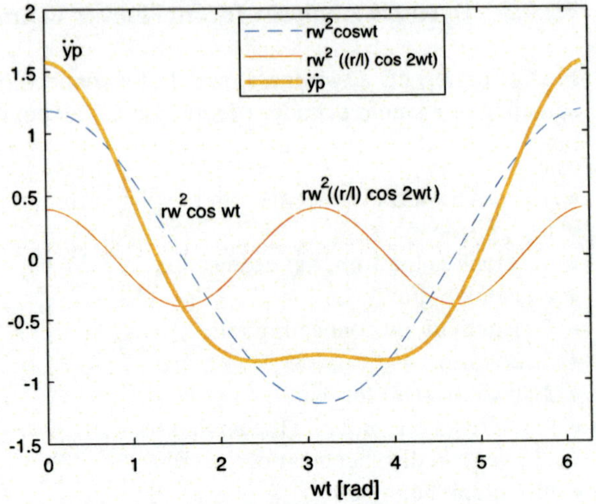

Fig. 7.8 The piston acceleration as a function of the crank angle for r/l = 0.33

The velocity and acceleration follow from the displacement by differentiation:

$$\dot{y}_p = r\omega\left[\sin(\omega t) + \frac{r}{2\ell}\sin(2\omega t)\right] \tag{7.14}$$

$$\ddot{y}_p = r\omega^2\left[\cos(\omega t) + \frac{r}{\ell}\cos(2\omega t)\right] \tag{7.15}$$

Figure 7.8 shows the piston mass vertical (y direction) acceleration, where $\lambda = 0.33$. Consequently the vertical inertial force had the same shape. Having determined the dynamic behaviour of the piston, the next step corresponds to the rotating pieces of the crank. In this case, the inertial force applied to the centre of gravity of the crank, moved to the bolt; however, in this operation, it was inversely proportional to the distance from the centre of the crankshaft. Thus, the entire mass of the crank pieces is replaced by a single mass m_c at the crank bolt with the connecting rod. From Fig. 7.7, the vertical displacement is given by:

$$y_c = \ell + r\left[1 - \cos(\omega t)\right] \tag{7.16}$$

Then, the vertical components of velocity and acceleration are

$$\dot{y}_c = r\omega\sin(\omega t) \tag{7.17}$$

$$\ddot{y}_c = r\omega^2\cos(\omega t) \tag{7.18}$$

Finally the horizontal components of velocity plus acceleration are

$$\dot{x}_c = -r\omega\cos(\omega t) \tag{7.19}$$

Table 7.1 Engine mass and damper parameters. Corresponding undamped natural frequency

Symbol	Quantity	Value
m	Engine mass	65 Kgr
E	Teflon Young Module	0.4×10^9 Pa
R_e	Damper external radius	15×10^{-3} m
R_i	Damper internal radius	5×10^{-3} m
ℓ	Damper length	10^{-3} m
A	Damper section $\pi \left[R_e^2 - R_i^2\right]$	$6.2832 \times 10^{-4} m^2$
$K_i = AE/\ell$	Damper spring constant	$2.5133 \times 10^7 \; [N/m]$
$K_e = 4K_i$	Four dampers equivalent spring	$1.0053 \times 10^8 \; [N/m]$
ω_n	Undamped frequency	$1.2436 \times 10^3 \; [rad/s]$

$$\ddot{x}_c = r\omega^2 \sin(\omega t) \tag{7.20}$$

Again, the corresponding inertial forces had the same shape and could be projected onto the four-bar linkage plane, as shown in Fig. 7.8. The engine mass, spring and damper properties are listed in Table 7.1, according to the initial dynamic model proposed in Fig. 7.3. The simulation responses of this model to the input of inertial forces are shown in the overall Fig. 7.9. The amplitude of the vibration that undergoes the revolute joint O_4 can vary greatly, especially as a function of the percentage of engine inertial force imbalance, as demonstrated in Fig. 7.10. The dynamic balancing of the engine crankshaft is never perfect and can significantly influence the resulting vibration. For example, the variations in the amplitude of the vibration are shown in Fig. 7.10b. If the percentage of inertial force imbalance is not large enough, let us accept 0.5 %; then, the resulting amplitude of vibration can slightly exceed $\Delta = 2$ [mm]. This result was extracted from Fig. 7.9b, which represents the underdamped response of the model depicted in Fig. 7.3, the parameters of which are listed in Table 7.1. Furthermore, a perfect crankshaft balance can result in an excessively optimistic assumption regarding the probabilistic engineering. A small increase in the crankshaft imbalance resulted in a substantial increase in the amplitude of the vibration, as demonstrated in Fig. 7.10.

Section two describes the significance of achieving the correct range of variation of φ_4 when designing the four-bar linkage. However, this range is not the only important parameter; the selection of a well-suited rocker arm angular velocity can also have a large impact on clutch performance. Note that the velocity ratio between the output and input links of the four-bar linkage can be viewed as instantaneously equivalent to the speed ratio between two gears in contact at the instant centre $I_{4/2}$, as shown in Fig. 7.5 [20]. As stated in Sect. 7.2 such a speed ratio is given by

$$\frac{\omega_4}{\omega_2} = \frac{O_2A}{O_4B} \cdot \frac{\sin(\nu)}{\sin(\mu)} \tag{7.21}$$

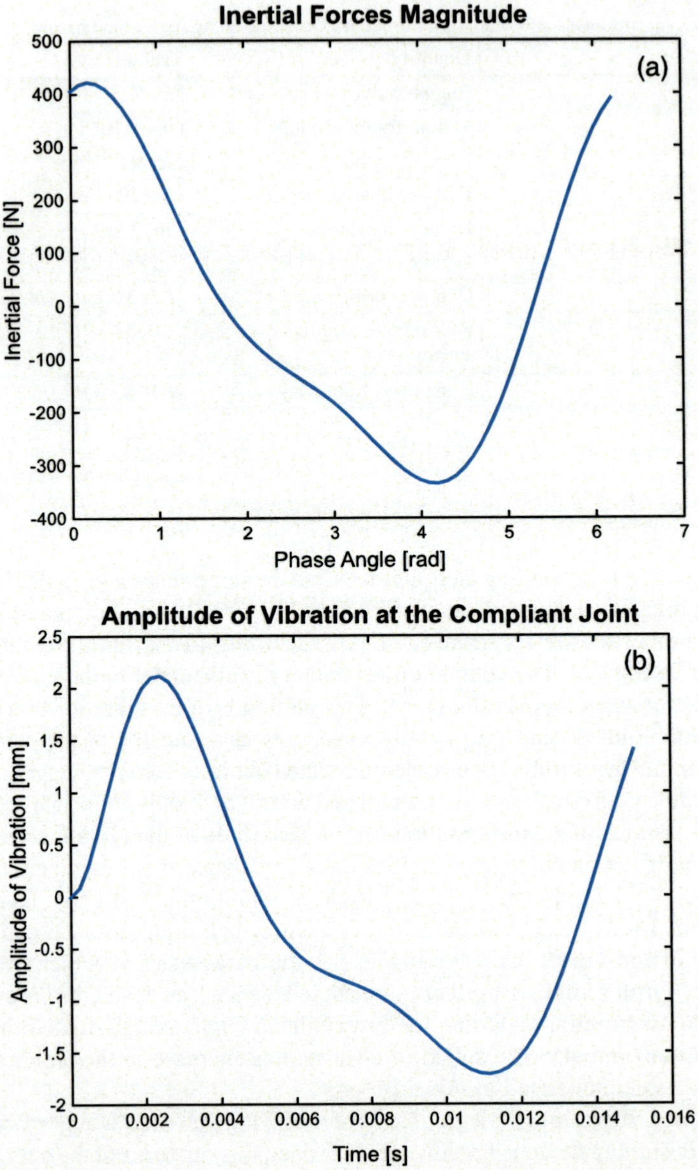

Fig. 7.9 **a** The magnitude of the inertial forces projected on the mechanism plane. **b** The amplitude of Vibration. Conditions: engine speed 418 [rad/s] (3991.6 rpm), system natural frequency: 198 [Hz], damping ratio ζ: 0.5

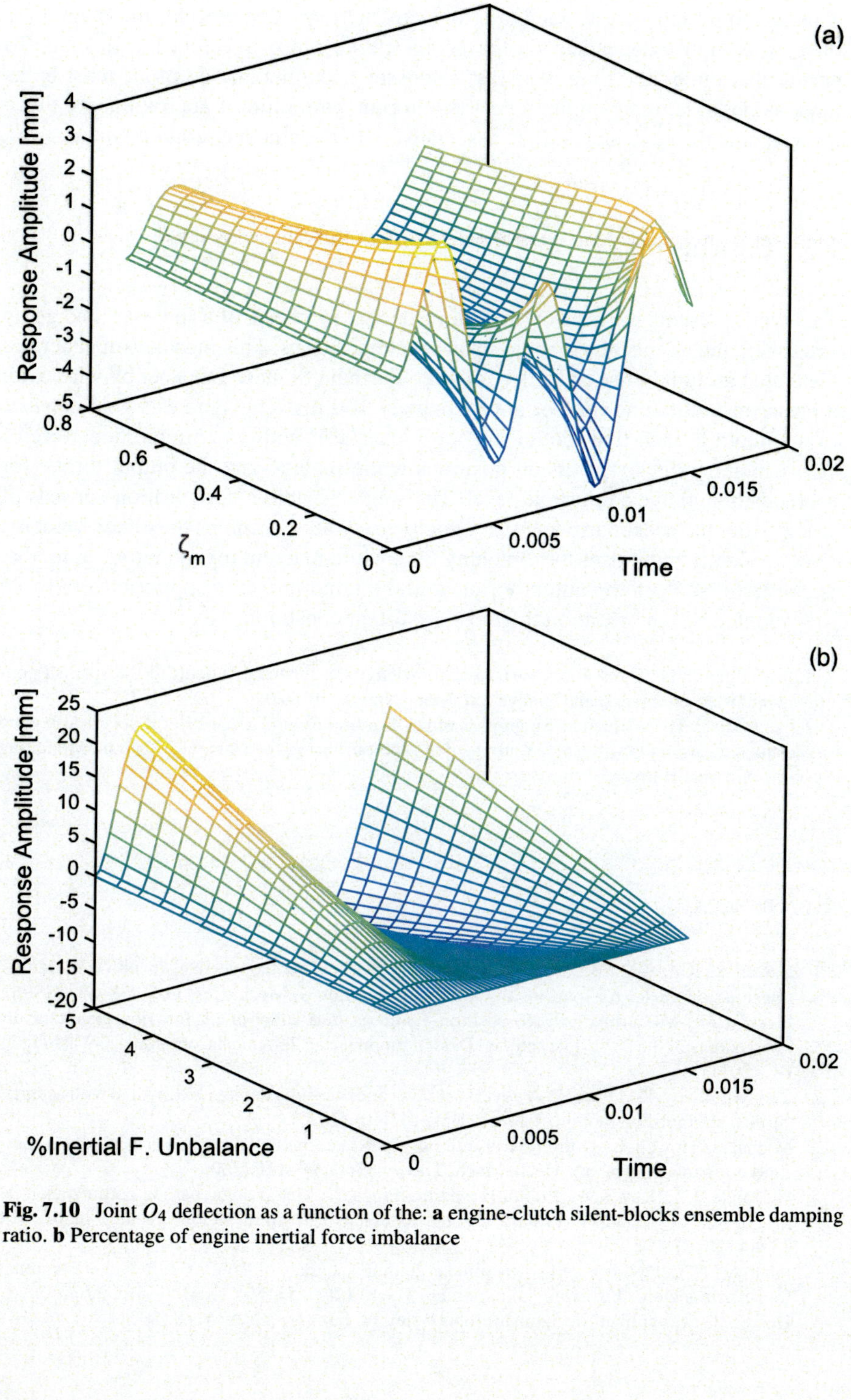

Fig. 7.10 Joint O_4 deflection as a function of the: **a** engine-clutch silent-blocks ensemble damping ratio. **b** Percentage of engine inertial force imbalance

This was repeated here to facilitate this explanation. As expected, the magnitude of ω_4 was highly dependent on the geometric design of the four-bar linkage. To ensure that the clutch time is as fast as possible, the magnitude of ω_4 must be as large as possible, maintaining a desirable torque gain ratio of approximately twice to overcome the resistance torque associated with the clutch operation.

7.3 Conclusion

This work presents the near-optimal dimensional synthesis of a four-bar linkage by discussing the effects of a pivot point at toggle positions. This mechanism generates the clutch motion of the University of Vigo Formula Student Car steer by wire, with a prescribed clutching time of approximately 100 ms. The synthesis goal consists of a smooth branch-free motion for the rocker arm with an extreme departure of 12°, which entails six precision positions while maintaining the output torque for each position above a threshold level. The single-objective optimisation consists of maximising the rocker arm angular velocity ω_4, which becomes the clutch velocity, or equivalently minimises the clutching time while maintaining the torque gain, that is, the ratio between the output torque and the input torque is approximately twice to guarantee clutch operation the most important constraint.

Acknowledgements The authors wish to acknowledge the financial support of the spanish goberment ministry of science and innovation Grant Number PID2020-116984RB-C21. We would also like to thank Dr. William E. Singhose and Dr. Tarunraj Singh for his helpful perspectives and suggestions regarding how to solve optimisation problems using the General Algebraic Modelling Systems (GAMS) software.

References

1. Bawab, S., Kinzel, G., Waldron, K.: Rectified synthesis of six-bar mechanism with well-defined transmission angles for four-position motion generation. J. Mech. Des. **118**, 377–384 (1996)
2. Plecnik, M., McCarthy, J.: Five position synthesis of a slider-crank function generator. In: Proceedings of the 2011 International Design Engineering Technical Conferences, Washington DC (2011)
3. Cao, W., Zhang, D., Ding, H.: A novel two-layer and two-loop deployable linkage with accurate vertical straight-line motion. J. Mech. Des. **142**(10) (2020)
4. Midha, A., Howell, L., Norton, W.: Limit positions of compliant mechanism using the pseudo-rigid body model concept. Mech. Mach. Theory **35**(1), 99–115 (2000)
5. Halzague, J.A., Key, R.A., Rubio, H., García-Prada, J.: Metodología para la optimización del mecanismo de suspensión trasera de una motocicleta de competición. Técnica Indus. **314**, 38–44 (2016)
6. Pavlovic, N.T., Pavlovic, N.D.: Rectilinear guiding accuracy of Roberts-Tchebichef compliant four-bar linkage with silicone joints. Mech. Mach. **12**(4), 53–56 (2004). (in Bulgarian)
7. Howell, L.: Compliant Mechanism. John Wiley I& Sons Inc., New York (2001)

8. Midha, A., Norton, T., Howell, L.: On the nomenclature, classification and abstractions of compliant mechanisms. J. Mech. Des. **116**(1), 270–279 (1994)
9. Howell, L., Midha, A.: A method for the design of compliant mechanisms with small-length flexural pivots. J. Mech. Des. **116**(1), 280–290 (1994)
10. Ananthasuresh, G., Kota, S.: Designing compliant mechanisms. J. Mech. Des. **117**(11), 93–96 (1995)
11. Baottcher, F., Christen, G., Pferkorn, H.: Structure and Function of Joints and Compliant Mechanism, pp. 30–35. Friedrich-Schiller Universitat Jena; Technische Universitat Jena; Technische Universitat Ilmenau; Shaker VerlagShaker Verlag, Aachen (2001)
12. Angeles, J., Alivizatoss, A., Akhras, R.: An unconstrained nonlinear least-square method of optimization of RRRR planar path generator. Mech. Mach. Theory **23**, 343–353 (1988). https://doi.org/10.1016/0094-114X(88)90048-1
13. Adrian, V., Yannou, B.: Dimensional synthesis of planar mechanisms using neural networks: application to path generator linkages. Mech. Mach. Theory **36**, 299–310 (2001). https://doi.org/10.1016/S0094-114X(00)00037-9
14. Urizar, M., Oyerro, A., Amezua, E.: Optimum Dimensional Synthesis using GIMSYNT Software. In: Uhl, T. (eds.) Vol. 73, pp. 785–754. Krakow (2019)
15. Kawamoto, A., Bendse, M., Sigmund, O.: Articulated mechanism design with a degree of freedom constraint. Int. J. Numer. Methods Eng. **61**, 1520–1545 (2004). https://doi.org/10.1002/nme.1119
16. Stolpe, M., Kawamoto, A.: Design of planar articulated mechanisms using branch and bound. Math. Program **103**, 357–397 (2005). https://doi.org/10.1007/s10107-005-0586-3
17. Fox, R., Willmert, K.: Optimum design of curve-generating linkages with inequality constraints. J. Eng. Ind. **89**, 144–151 (1967). https://doi.org/10.1115/1.3609986
18. Villegas, C., Husing, M., Corves, B.: Dynamic Synthesis of a Crank-Rocker Mechanism Minimizing its Joint-Forces. In: Uhl, T. (eds.) Vol. 73, pp. 3167–3176. Krakow (2019)
19. Iwatsuki, N.: Mobile Machines Easily Assembled with Planar link Parts for Designing and Prototyping Practice to Inspire Junior/Senior High School Students. In: Uhl, T. (eds.) Vol. 73, pp. 853–862. Krakow (2019)
20. McCarthy, J.M., Soh, G.S.: Springer, New York (2010)
21. Ceccarelli, M.: Mechanism schemes in teaching: a historical overview. J. Mech. Des. **120**, 533–541 (2012). https://doi.org/10.1115/1.2829311
22. Flores, P., Seabra, E.: Dynamics of Planar Multibody Systems. VDM Verlag Dr. Muller GmbH and Co KG, Saarbrucken (2010)
23. Nikravesh, P.: Planar Multibody Dynamics: Formulation Programming and Applications. CRC Press, Boca Raton (2008)
24. de Jalón, J.G., Bayo, E.: Kinematic and Dynamic Simulation of Multibody Systems: The Real-Time challenge. Springer-Verlag, New-York (1994)
25. Haugh, E.: Computer Aided Kinematics and Dynamics of Mechanical Systems. Allyn and Bacon Series in Engineering, Massachusetts (1989)
26. Hartog, J.P.D.: Mechanical Vibrations, 4th edn. McGraw Hill, New York (1956)

Chapter 8
Dimensional Synthesis Solution of a Compliant Clutch Mechanism for a Formula Student Car (II)

Gerardo Peláez, Gustavo Peláez, Higinio Rubio, and Alejandro Bustos

Abstract A physical four-bar linkage prototype was synthesised in which one pivot revolute joint was linked to the engine assembly of a Formula Student Car. Thus, the mechanism frame became a flexible link. The mechanism has a large torque gain, as with a low torque at the input crank supplied by a servomotor steer by wire, and a high output torque is produced at the rocker arm. The entire system demonstrated good performance in terms of its reliability when operating the clutch, driving ergonomic operation using cams at the steering wheel and a well-suited clutching time when compared to the other solutions.

Keywords Compliant mechanism · Near-optimal · Dimensional synthesis · Steer by wire

8.1 Introduction

Numerical methods that obtain approximate solutions [1] becomes the more appealing option when more than five precision points are specified for the dimensional synthesis of a mechanism. To this end, numerical gradient-based methodologies are promising owing to their computational efficiency [2, 3]. In the following the General Algebraic modelling Systems (GAMS) software was selected. GAMS is particularly recommended, as we deal with a high number of nonlinear inequality constraints to capture the increase in mechanism mobility. Primary work in this field was conducted by Fox and Willmert [4]. The present work addresses both mechanism optimisation

G. Peláez (✉) · G. Peláez
University of Vigo, Vigo 36310, Spain
e-mail: gpelaez@uvigo.es

H. Rubio
University Carlos III de Madrid, Getafe, Spain

A. Bustos
UNED, Madrid, Spain

J. C. García Prada et al. (eds.), *Trends in Educational Activity in the Field of Mechanism and Machine Theory (2018–2022)*, Mechanisms and Machine Science 128,
https://doi.org/10.1007/978-3-031-25730-8_8

and assembly prototyping of the near-optimal solution found for the four-bar pseudo-rigid linkage. For a more in-depth review of past and present mechanisms, readers are referred to [5, 6].

8.2 Synthesis of the Four-Bar Linkage

The dimensional synthesis of the mechanism depicted in Fig. 8.1 consists in minimizing the operation time while keeping the output torque above a threshold value, this can be generalised analytically as follows:

$$minimize\rightarrow -\omega_4 = -\left\{\frac{O_2A}{O_4B}\cdot\frac{sin(\nu^*)}{sin(\mu^*)}\cdot\omega_2\right\} \tag{8.1}$$

Therefore the final goal consist in minimize $-\omega_4$, where ν^* and μ^* are the worst values for both angles when considering the speed ratio, in our case $\nu^* = 30°$ and $\mu^* = 60$. To summarise, the four-bar linkage is designed by satisfying Eq. 8.1 as the objective or goal function, subject to the inequality algebraic closed loop plus torque gain constraint equations given by:

$$\ell_2\cdot\cos(\varphi_2(i)) + \ell_3\cdot\cos(\varphi_3(i) - (\ell_4\cdot\cos(\varphi_4(i)) + d) \geqslant -\Delta \tag{8.2}$$

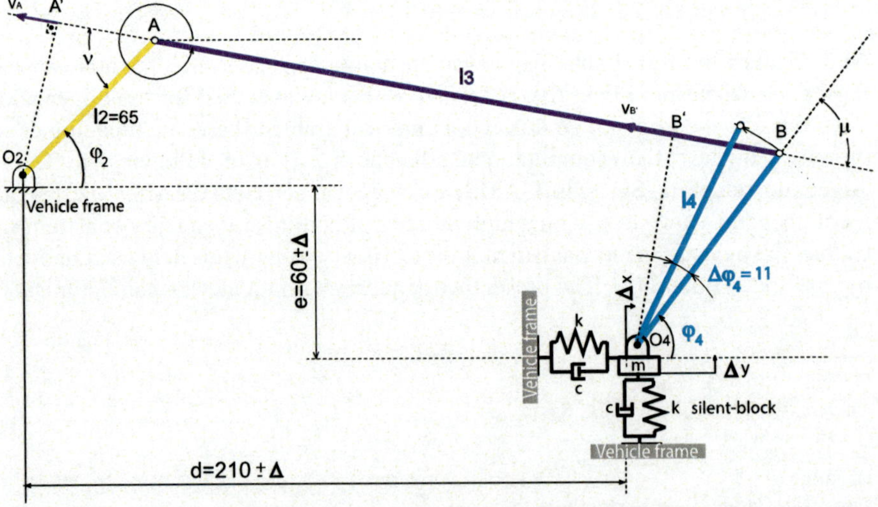

Fig. 8.1 The mechanism plus the effective links. The link lengths and the angles to be estimated are φ_3, φ_4, ℓ_3, ℓ_4. Engine amplitude of vibration Δ. Conditions: No mismatch between the axial parameters

Table 8.1 Relationship between φ_2, φ_3 , φ_4, ν, μ , ω_4 and T_4

Position	φ_2	φ_4	φ_3	ν	μ	ω_4 [rad/s]	T_4 [N · m]
1	33	44.15	355.8526	37.1474	48.2974	1.8602	22.5553
2	36	46.17	354.8958	41.1043	51.2743	1.9381	21.6484
3	39	48.25	355.8927	43.1073	52.3573	1.9848	21.1387
4	42	50.47	354.9473	47.0527	55.5247	2.0421	20.5457
5	45	52.54	354.9474	50.0527	57.5927	2.0884	20.0903
6	48	54.61	354.9531	53.0469	59.6549	2.1297	19.7012

$$\ell_2 \cdot \cos(\varphi_2(i)) + \ell_3 \cdot \cos(\varphi_3(i) - (\ell_4 \cdot \cos(\varphi_4(i)) + d) \leqslant \Delta \tag{8.3}$$

$$e + \ell_2 \cdot \sin(\varphi_2(i)) + \ell_3 \cdot \sin(\varphi_3(i)) - (\ell_4 \cdot \sin(\varphi_4(i))) \geqslant -\Delta \tag{8.4}$$

$$e + \ell_2 \cdot \sin(\varphi_2(i)) + \ell_3 \cdot \sin(\varphi_3(i)) - (\ell_4 \cdot \sin(\varphi_4(i))) \leqslant \Delta \tag{8.5}$$

$$T_4 - T_2 \cdot \frac{\ell_4}{\ell_2} \cdot \frac{\sin(\varphi_4(i) - (\varphi_3(i) - 2 \cdot \pi))}{\sin(\varphi_2(i) - (\varphi_3(i) - 2 \cdot \pi))} \leqslant 0 \tag{8.6}$$

The variables to be estimated are ℓ_3 and ℓ_4, and the values of $\varphi_3(i)$ for each of the six positions, that is, varying the index i from one to six. From this perspective the General Algebraic Modelling System (GAMS) was used to solve the nonlinear optimisation problem. CONOPT was selected as the gradient-based numerical method. The outcomes corresponding to the feasible solutions are listed in Table 8.1. To satisfy the 18 constraints, the optimal solution yields the following longitudes corresponding to the connecting rod plus the rocker arm:

$$\begin{bmatrix} \ell_3 \\ \ell_4 \end{bmatrix} = \begin{bmatrix} 230 \\ 110 \end{bmatrix} \tag{8.7}$$

These results were fed into SAM 7 software ©ARTAS to simulate the kinematics and dynamics of the optimal four-bar linkage. For all the considered cases, as demonstrated in Fig. 8.2 and the Table 8.1, the actual positions accomplish the requirements, maintaining a smooth change for the angle φ_3, which is a mechanism of the type crank-rocker arm without a branch. The rocker arm angular velocity value ω_4 is close to 2 rad/s, maintaining the supplied output torque at approximately T_4 = 20 N·m, which is nearly twice the input torque value of T_2 = 10.78 N · m. In addition, the clutching time value for the four-bar parallelogram last for 77.04 milliseconds (ms), whereas for the optimal four-bar linkage synthesised, the clutching time becomes approximately 3 ms larger for the range of φ_4 [44.15 54.61] degrees. The clutching time increased only slightly, as shown in Fig. 8.3. This is the price necessary to achieve a high output torque T_4 twice the input torque T_2.

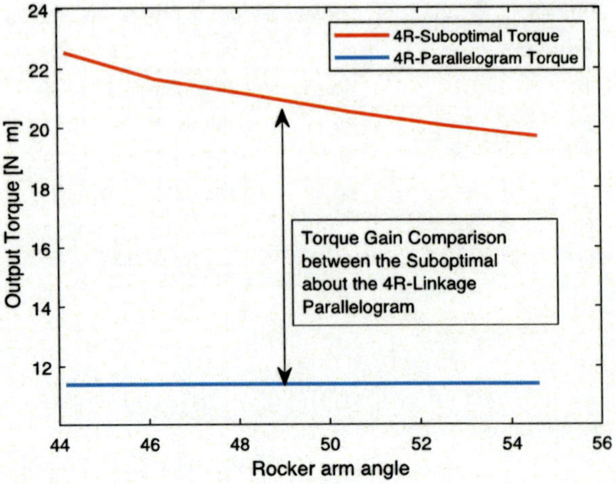

Fig. 8.2 Torque gain comparison between the optimised about the 4r-linkage parallelogram mechanisms simulation responses. Conditions: ω_2 = 3.892 rad/s

8.2.1 Performance Indices

To reinforce these results, performance indices were defined to quantify and compare the optimal four-bar linkage with the standard four-bar parallelogram linkage given by $\ell_2 = 65$ $\ell_3 = 218$ and $\ell_4 = 65$. Figure 8.2 compares the available rocker arm output torque T_4 faced with the crank input torque supplied by the servomotor T_2 assumed to be constant. The vertical scale was shifted upward so that the output torque in each position was clearly visible. The optimal mechanism used to generate Fig. 8.2 is given by $\ell_2 = 65$, $\ell_3 = 230$, and $\ell_4 = 110$ [mm]. As shown in Fig. 8.2, it is quite effective to amplify the input torque level. Thus, the first considered performance index, the torque gain TG throughput mean value, becomes

$$TG = \frac{T_4^{mean}}{T_2^{const}} = \frac{20.9466}{10.78} = 1.94 \tag{8.8}$$

Even though, as expected, the mean value of the angular velocity ω_4 corresponding to the rocker arm is approximately 2 [rad/s] according to the simulation results depicted in Table 8.1, it has a lower value than the input angular velocity of the crank ω_2=3.89 [rad/s] supplied by the servomotor. This is in agreement with Kennedy's theorem, as shown in Fig. 8.4.

$$\frac{\omega_4}{\omega_2^{mean}} = \frac{O_2 I_{4/2}}{O_4 I_{4/2}} = 0.5141 \tag{8.9}$$

As stated above, the only drawback of the optimal four-bar linkage is that it requires a slightly longer clutching time than the single-parallelogram four-bar linkage shown in Fig. 8.3. Nonetheless, 80 ms accomplishes the prescribed clutching time limit of

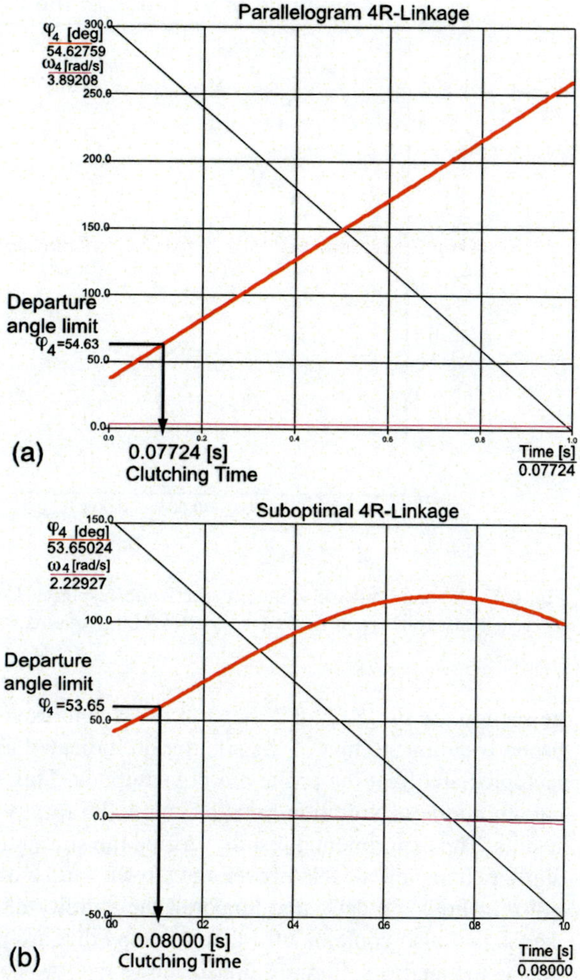

Fig. 8.3 Comparison of **a** Clutch response actuated by the four-bar parallelogram and **b** the optimised four-bar linkage. Conditions: Crank angular velocity $\omega_2 = 3.89$ rad/s

100 ms. We understand that we cannot obtain something from nothing. In this study, we attempted to obtain a high output torque T_4 that is twice the input torque T_2 from very little.

8.3 Experimental Results

Figure 8.5 shows the the Formula Student Car used to perform experiments at the Montmelo Circuit. An analog distance probe, with reference XS4P12AB120 Telemecanique, capable of measuring distance in the range [0.4–4] mm was located close to the engine surface at a distance of around 1.5 mm as shown in Fig. 8.6a. As can

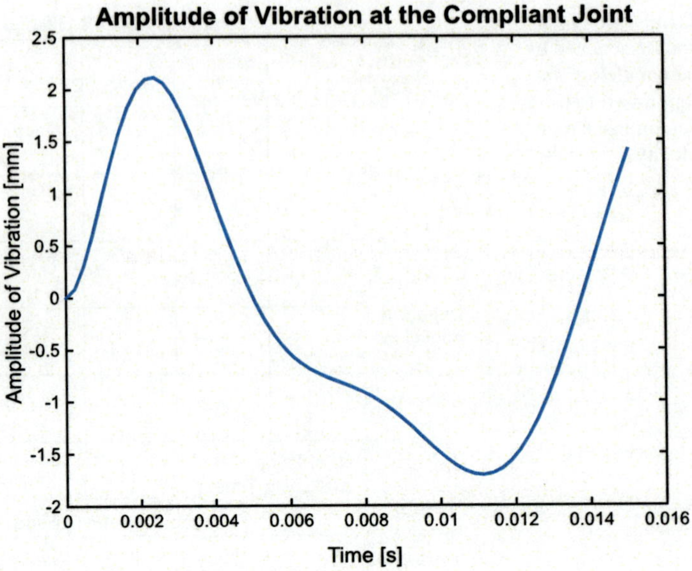

Fig. 8.4 The amplitude of vibration simulation response. Conditions: engine speed 418 [rad/s] (3991.6 rpm), system natural frequency: 198 [Hz], damping ratio ζ: 0.5

be seen in the Fig. 8.6b, the shape of the signal provided by the downward pointing probe is similar to that of the simulation indicated above in Fig. 8.4. In addition, it is appreciated that the probe did not saturates. This indicates that the engine vertical amplitude of vibration remains below 2.5 mm, which reinforces the simulation results. The experimental results verified the simulation data because they exhibited similar characteristics. However, the largest source of error, is likely to be the sensor noise in Fig. 8.6b data, that hindered the validity of the measured dynamic signal. The signal also contains offsets and jumps that are processed by low-pass filtering during the analysis. Figure 8.6 also shows the low-pass-band filtered signal(cian).

To verify the smooth motion of the connecting rod, simulation was conducted using the aforementioned two-dimensional SAM 7 kinematics and dynamics simulator © ARTAS. Figure 8.7a shows the path of the revolute joint B branch free that corresponds to a smooth motion of the rocker arm. The setup, shown in Fig. 8.7a, was constructed to verify this behaviour. The Fig. 8.7b shows the actual manufactured mechanism for which the length of the links corresponds to those calculated according to the previous dimensional synthesis. The experimental clutching time behaves below 80 milliseconds (ms) according to the Formula Student Car response. While the output torque supplied by the mechanism was able to counteract the resistant torque of the clutch for all situations.

Fig. 8.5 The formula student car in the montmelo circuit

8.4 Conclusion

This work presents the near-optimal dimensional synthesis of a four-bar linkage. The six precision positions, coming from an allowance of the rocker arm rotation below 12°, allow the set up of twelve closed-loop algebraic equations that are somewhat relaxed to inequality constraints because the revolute joint O_4 is linked to the engine assembly that undergoes mechanical vibrations. The amplitude of the vibration caused by the engine inertial forces was estimated by simulation, resulting in less than $\Delta = 2$ mm, which was corroborated by experimental measurements. Using Δ as the limit for inequality constraints helps quickly obtain an optimal solution for the stated optimisation problem. Exploitation of flatness near optimal solutions is useful for mechanism synthesis. The dimensional synthesis problem is solved using numerical techniques, specifically the reliable General Algebraic Modelling Systems (GAMS) software. The simulation of the resulting four-bar linkage determines that fulfils the imposed constraints. An output torque of approximately 20 [N · m] was achieved, whereas the clutching time of 80 ms was only 3 ms longer than that corresponding to a four-bar linkage parallelogram. A probabilistic analysis dealing with the unbalanced percentage of engine inertial forces was developed to ensure

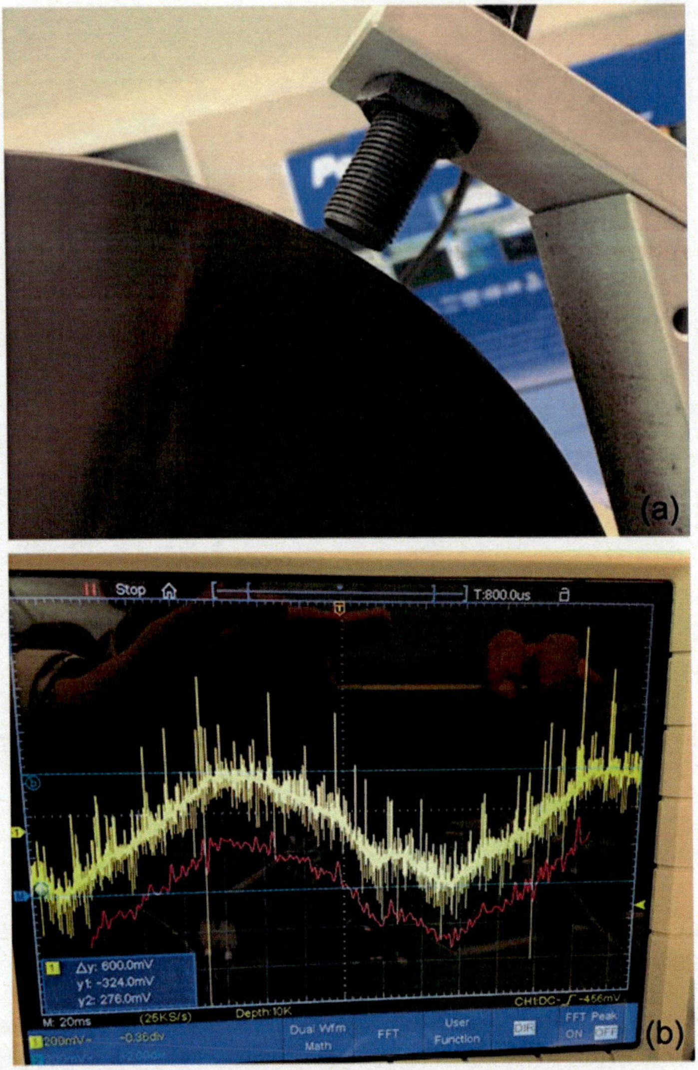

Fig. 8.6 **a** Probe layout. **b** Experimental dynamic signal measured (yellow) and Low-pass-band filtered signal(cian), 600 [mV] signal amplitude. Engine condition: 1050 [rpm]

the reliability of the imposed inequality constraints. Although somewhat limited in scope, the experimental results of the Formula Student Car shown in Fig. 8.5 reinforce the notion that the synthesised four-bar linkage has good performance regarding the driving ergonomic operation using cams at the steering wheel and a well-suited clutching time when compared to the other solutions.

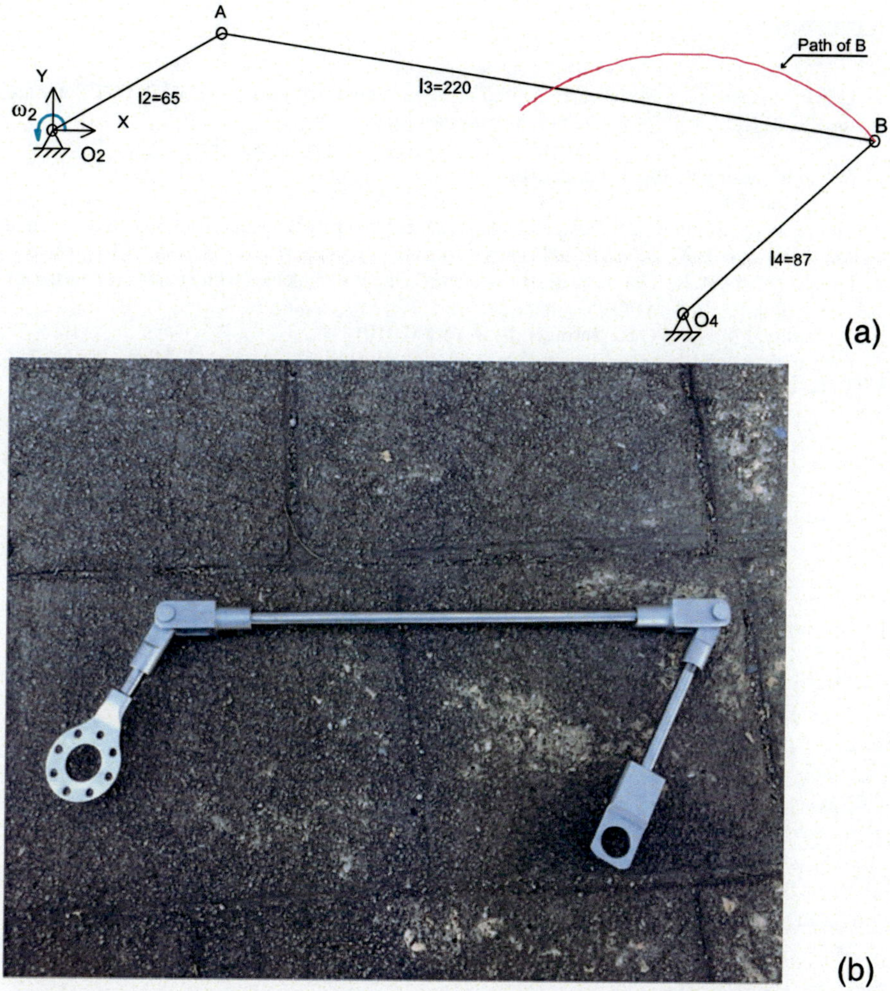

Fig. 8.7 The actual mechanism. Conditions: **a** smooth motion of the connecting rod **b** the manufactured mechanism

Acknowledgements The authors wish to acknowledge the financial support of the spanish goberment ministry of science and innovation Grant Number PID2020-116984RB-C21. We would also like to thank Dr. William E. Singhose and Dr. Tarunraj Singh for his helpful perspectives and suggestions regarding how to solve optimisation problems using the General Algebraic Modelling Systems (GAMS) software.

References

1. Urizar, M., Oyerro, A., Amezua, E.: Optimum dimensional synthesis using GIMSYNT Software. In: Uhl, T. (Ed.), vol. 73, pp. 785–754. Krakow (2019)
2. Kawamoto, A., Bendse, M., Sigmund, O.: Articulated mechanism design with a degree of freedom constraint. Int. J. Numer. Methods Eng. **61**, 1520–1545 (2004). https://doi.org/10.1002/nme.1119
3. Stolpe, M., Kawamoto, A.: Design of planar articulated mechanisms using branch and bound. Math. Program **103**, 357–397 (2005). https://doi.org/10.1007/s10107-005-0586-3
4. Fox, R., Willmert, K.: Optimum design of curve-generating linkages with inequality constraints. J. Eng. Ind. **89**, 144–151 (1967). https://doi.org/10.1115/1.3609986
5. McCarthy, J.M., Soh, G.S.: Springer, New York (2010)
6. Ceccarelli, M.: Mechanism schemes in teaching: a historical overview. J. Mech. Design **120**, 533–541 (2012). https://doi.org/10.1115/1.2829311

Part II
Mechanism and Machine Science in the Engineering Program: Experiences and New Trends

Chapter 9
Methodological Approach for Interdisciplinary Teaching of Machines and Mechanism in Real Contexts

José Antonio Hernández-Torres, Juan Torreglosa, María Reyes Sánchez-Herrera, Ángel Mariano Rodríguez-Pérez, Julio José Caparrós Mancera, and Anna Cislowska

Abstract Today's world is full challenges and uncertainties. In the context of continuous changes and evolutions, future engineers need to be able to access, comprehend, assess, synthesize, and apply viewpoints and knowledge from different domains. Since one of the main problems new generations have moved to environmentally related task, such as clean water solution, or energy generation and management. The experience obtained by researcher while working in interdisciplinary international projects like Reffect Africa, where the lack of access to basic results is tackled must be transferred to students. In this context, this work presents a methodology to approach machines and mechanism theories to electrical engineers. Likewise, another methodology to flip the previous target by using machines design is introduced. This work focusses on discover what learning process stimulate the development of technical knowledge and acquisition to develop new trends and teaching plans.

Keywords Machines · Mechanisms · Gasifier · Electrical engineering

9.1 Introduction

Today's challenges are complex and often open ended and not defined [1]. These difficulties transcend the typical perception of the duties and obligations of engineers [2]. They require a certain kind of engineer that is sociable and can operate both inside and outside the confines of their own field [3]. Future engineers therefore need to be able to access, comprehend, assess, synthesize, and apply viewpoints and knowledge from domains other than their own.

J. A. Hernández-Torres (✉) · J. Torreglosa · M. R. Sánchez-Herrera · Á. M. Rodríguez-Pérez · J. J. Caparrós Mancera · A. Cislowska
University of Huelva, Avenida 3 de Marzo S/N, Huelva, Spain
e-mail: joseantonio.hernandez@dimme.uhu.es

J. C. García Prada et al. (eds.), *Trends in Educational Activity in the Field of Mechanism and Machine Theory (2018–2022)*, Mechanisms and Machine Science 128,
https://doi.org/10.1007/978-3-031-25730-8_9

According to previous research, interdisciplinarity has traditionally been seen as linking various technical areas, with only a few examples of larger-scale partnerships. investigations into fresh, successful methods for integrating interdisciplinarity into engineering education are often impeded by the historically discipline-oriented nature of academia [4]. Since official study programs follow national and international standards, it is a duty of the academics to develop new strategies and methodologies in the regulatory framework. Overcoming this can open up new opportunities to get our students to acquire interdisciplinary skills for their future professions and professional development [5].

Related to this context, one of the most relevant situations where collaboration between engineers from different fields is required is the access to electricity and energy. Even in the twenty-first century, a significant portion of the global population still does not have access to the minimum amount of energy they require. Worldwide waste from agriculture and agri-food processing is enormous, and these industries also use a lot of electrical and, or thermal energy. Typically, these wastes are not valorized, which has detrimental effects on the global sustainability of the agri-food business [6]. Researchers from several nations conducted numerous experimental studies using biomass obtained from producer gas as a substitute for fuel for internal combustion engines. These findings demonstrated that, when using producer gas, thermal efficiency was comparable to that of diesel with just minor power loss.

In attempt to address these difficulties, the project Reffect Africa suggests designing and constructing a cutting-edge, dependable, and unique sustainable energy solution based on the valorization of biomass wastes from the food and agricultural industries utilizing biomass gasification. Three full-scale demonstrations will be built in Morocco, Ghana, and South Africa to account for both urbanized and rural locations in Africa, on- and off-grid solutions, as well as different socioeconomic backgrounds. This project addresses the development of renewable energy sources, off-grid and on-grid community solutions, and their integration into the existing energy infrastructure. It will consider the generation of renewable energy, the transmission, and the use of storage systems [7]. Combining the engineering teaching processes and the knowledge and experience provided by projects like Reffect Africa, the objective of this work to develop a methodology to deepen in students' knowledge in the fields of mechanical, electricity and energy engineering.

9.2 Methodology

This work implements a quantitative study while analyzing the application real cases of study in different engineering courses. Unlike most previous studies [8–10], this work was created during a two-year period concurrently in various engineering degrees. Comparing the various degrees of transversal knowledge acquisition with a focus on how much more knowledge graduates of a given bachelor's program have about the various engineering specialties is the goal. Thus, it allows for the monitoring of the development of both electrical and energy engineering bachelor

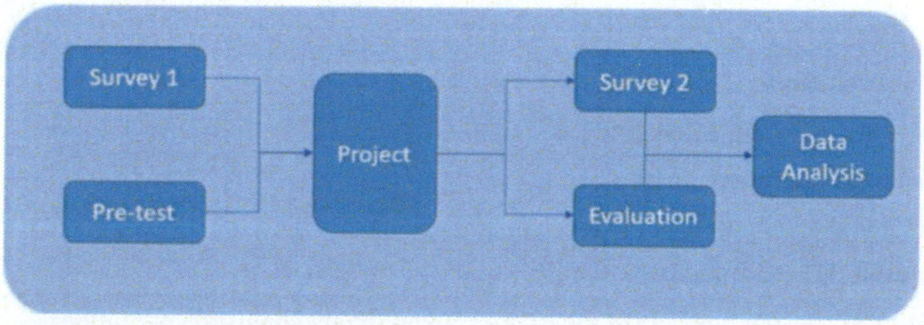

Fig. 9.1 Methodology flowchart

students' and mechanical engineering students' understanding of machinery used in electricity generation.

The proposed general methodology is subdivided into 4 stages, as shown in Fig. 9.1. Each of these stages take place during the year, and it repeats during the whole duration of the study. During the first practical class students are requested to perform a little task to show the initial level of the skills required for the subject. Additionally, during the first class, students fulfil a survey to provide a self-evaluation of their skills and the relevance of the content of the subject. After the first part, along the semester, and in parallel with the lectures, students have to develop the study and 3D model of a mechanism. This work is performed in groups of 2 up to 3 members. At the end of the semester, and before the marks get public, the students have to fulfil the same survey they did at the beginning of the semester. Finally, all the data are gathered and analyzed.

Once the general methodology is known, it is necessary to detail it. The first element is the pre-test. This activity consists on the generical analysis and a 3D model of a thrust crank mechanism. The election is based on the simplicity of this mechanism and its common usage as example during the subject. Students have to develop a simple report regarding the elements and physics as well as the construction of the mechanism.

The survey, which is completed at the beginning and at the end of the semester, is the second component of the first stage. By conducting the survey shown in Table 9.1, the didactic experiences of this work are assessed from the students' point of view. The concepts that are taught are used to evaluate various factors. the manner of instruction, the challenges pupils experience, how well students use their knowledge, and their level of satisfaction. On a scale of 1 to 5, where 1 represents the strongest disagreement and 5, the strongest agreement, each of the parameters is evaluated.

The key point of the methodology is the project that the students develop during the semester. Students are requested to develop the study and design of a mechanism. A list of mechanism is provided to the students, who are grouped in teams of two up to three people (due to the number of students in some of the courses it is necessary to set a low number of members per group to increase the sample of different works). Each group selects one mechanism from the list. Once a mechanism is selected it is not available for the rest of students. The task to perform consist in two main sections a technical report and a working 3D model. These are described below:

Table 9.1 Survey

Evaluation parameters				
1. The previous knowledge is helpful to complete the task				
2. The content is complex				
3. I understand the operation of the mechanism				
4. The software is user-friendly and intuitive				
5. The use of 3D design has helped me to understand the mechanism's operation				
6. My knowledge in machines and mechanisms theory and 3D design has increased				
7. The acquired knowledge has real utility				
8. The time to perform the task has been adequate				
9. I would like to continue learning about 3D design during engineering training				
10. I am satisfied with this experience				
Rating				
Completely disagree	Disagree	Neutral	Agree	Completely agree
1	2	3	4	5

- Technical Report: Students have to organize the following tasks
  - Contextualization of the usage of the mechanism.
  - Mechanical characterization.
  - Kinematic and dynamic analysis.
  - Annexes: Draws, Economical Study, Etcetera.
- 3D design: A real implementation of a working model of the mechanism. It should include as many details as possible.

To evaluate the different projects in the most impartial way, the evaluation of the results, follows a rubric specially designed for this purpose. Table 9.2 shows this rubric.

The key of the study is lies behind the fact that, while during the first year, the list of mechanisms is composed by random mechanism, it changes for the second year. The list of mechanisms has a biased offer during the second year. The list of mechanisms for electrical and energy engineering bachelor students is limited to mechanisms with a tight connection to the electrical field. For instance, a protective relay or an isolator switch, among others. Bachelor's degree holders in mechanical and electronic engineering, however, still have a selection of random mechanisms to select from. This is done in order to maintain a control group that will allow comparison of the outcomes of using this practice.

Table 9.2 Evaluation rubric

Criteria	Rating		
	1	2.5	5
1. Software	Use basic functions	Use advanced functions and shortcuts	Use 3D animation features
2. 3D design	The design contains some functional error	The design contains some non-functional error	Design is correct and functional
3. Mechanism design	The mechanism is simplified	The mechanism is complete	Adds additional functional parts and elements
4. Kinematic and dynamic analysis	Speed and acceleration are not properly calculated	Speed and acceleration are correct but dynamic analysis not	Kinematic and Dynamic problem correctly solved
5. Report results	Provides the report	Additionally, Provides design snapshots	Additionally, Provide design animations

9.3 Results

This section is composed by Fig. 9.2, which is a sample of the projects developed by the students and the Figs. 9.3 and 9.4 which represent the comparison of the results obtained in the pre-test and post-test in year one and year two.

Results of auto-evaluation and obtained marks are compared in Figs. 9.3 and 9.4. Where four different parameters are evaluated in order to classify the different skills the student will work on during the work. For each of the parameters, the values of auto-evaluation and marks are placed next to each other for a direct comparison. Additionally, these values are shown for all the four groups.

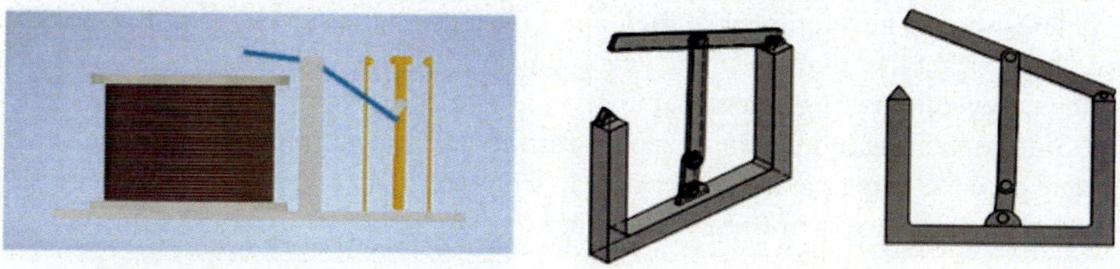

Fig. 9.2 Samples of 3D models developed by students

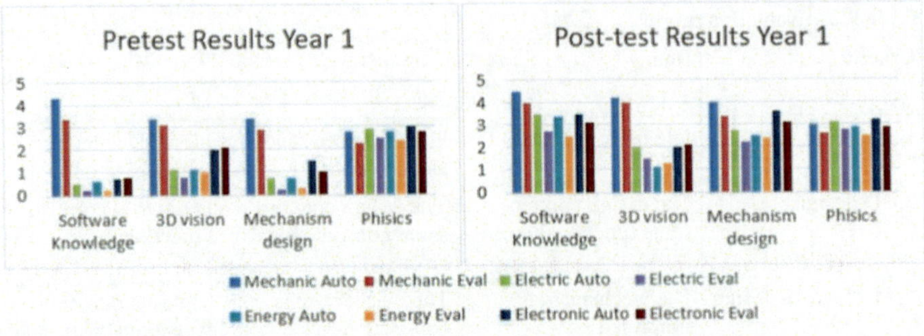

Fig. 9.3 Compared results of year 1

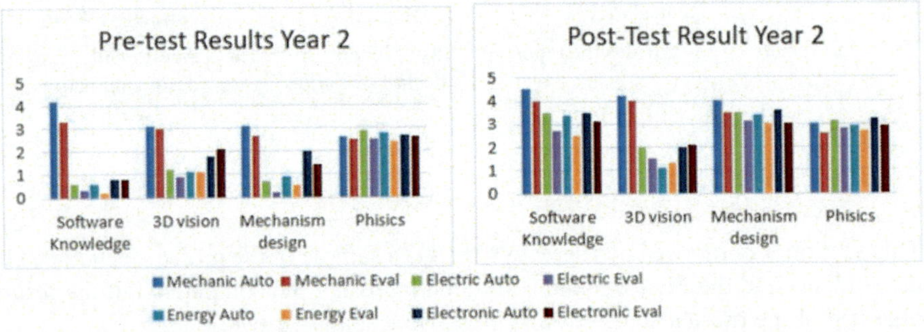

Fig. 9.4 Compared results of year 2

9.4 Discussion

Analyzing the data gathered from surveys and evaluations, it is possible to observe in a first approach that in both cases the progression at the end of the semester is relevant. While mechanical engineering students start with a higher knowledge and skills, this difference reduce notably. It is also necessary to consider that these students show a development of their abilities. During the second year, it is possible to observe that the increase in the development of the skills increase for the electric and energy engineering students.

Since obtaining more heterogeneous results reveals an optimization of the educational process, some parameters evolve much better than other. Thus, it is necessary to analyze the reasons of these differences.

9.5 Future Works

The knowledge and expertise obtained during the work of the researchers in the project Reffect Africa, allow to design new academic experiences. Working in real big-scale engineering projects also helps academic development [11]. In this way, based on the results of the previous experience, a new approach based on the same methodology is designed and currently on process.

Since the key point of the methodology is the project that the students develop during the semester, the new methodology varies not only in the context but mainly in the requested task. In this case, students have to develop the study and design of a whole system. The use of a gasifier to produce syngas to power electric generators. Electrical generators are commonly considered as a part of the electrical engineering field [12]. The objective is to approximate traditional electrical tasks to mechanical engineering students [13]. The task to perform consist in three main sections: A technical report, a deep mechanical study and a working 3D model. These are described below:

- Technical Report: Students have to summarize and organize the following tasks
  - Contextualization of the usage of the system.
  - Mechanical and electrical characterization.
  - Definition of the equipment.
  - Annexes: Draws, Economical Study, Etcetera.
- Deep mechanical study of one of the machines or mechanisms which compose the system
- 3D design: A real implementation of a working model of the mechanism. It should include as many details as possible.

Additionally, all the related elements (rubric, questionnaire and pre-test), are also modified to adapt the methodology to this approach. Finally, both methodologies will continue taking place in order to develop a longitudinal study.

9.6 Conclusion

The educational experience created for this work has demonstrated how using 3D design tools improves the engineering students' competencies. This enables them to apply theory in real-world settings. Particularly when designing mechanisms, the application of these techniques enables designers to more clearly visualize the corresponding functionality. It is intriguing to draw attention to the differences in how this experience is rated among engineering degrees. Their approach in mechanics was made easier by prior experience, and they excelled. In terms of their basic level, pupils are the ones who learned electronics to the greatest extent. There are significant obstacles and little success among energy students, but there is a lot of enthusiasm for continuing education in the field of 3D design application.

References

1. Gómez Puente, S., Van Eijck, M., Jochems, W.: Empirical validation of characteristics of design-based learning in higher education. Int. J. Eng. Educ. **29**(2), 491–503 (2013)
2. Vojak, B., Price, R., Griffin, A.: Corporate innovation. In: The Oxford Handbook of Interdisciplinarity. Oxford University Press, Oxford, UK, pp. 546–560 (2010)
3. Barut, M., Yildirim, M., Kilic, K.: Designing a global multi-disciplinary classroom: a learning experience in supply chain logistics management. Int. J. Eng. Educ. **22**(5), 1105–1114 (2006)
4. Van-den-Beemt, A., MacLeod, M., Van-der-Veen, J., Van-de-Ven, A., Van-Baalen, S., Klaassen, R., Boon, M.: Interdisciplinary engineering education: a review of vision, teaching, and support. J. Eng. Educ. **193**(3), 508–555 (2020)
5. Acuña-Peña, B.: Fórmulas para la innovación en la docencia universitaria. ACCI (Asociación Cultural y Científica Iberoamericana), Madrid (Spain) (2014)
6. Aguado, R., Vera, D., López-García, D., Torreglosa, J., Jurado, F.: Techno-economic assessment of a gasification plant for distributed cogeneration in the agrifood sector. Appl. Sci. **11**(2), 600 (2021)
7. Reffect Africa, "REFFECT AFRICA," (2022). https://www.reffect-africa.eu/#/about. Accessed Sept 10, 2022
8. Billings, C., Siddique, Z., Liu, Y.: Enhancement of mechanical engineering education with additive manufacturing projects. Eng. Educ. **9** (2020)
9. Chen, P., Zhang, T., Zhang, B., Luo, L.: Design and implementation of dynamic teaching model of diesel engine based on 3D printing technology. J. Phys. Conf. Ser. **2029**, 012092 (2021)
10. Caparrós-Mancera, J., Rodríguez-Pérez, A., Hernández-Torres, J.: Design and operational analysis of mechanical elements in different engineering. In: IEEE World Engineering Education Conference (EDUNINE), pp. 1–5 (2022)
11. Xeidakis, G.: Engineering education today: the need for basics or specialization. Eur. J. Eng. Educ. **19**(4), 485–501 (1994)
12. Malik, A., Mohapatra, S.: Biomass-based gasifiers for internal combustion (IC) engines—a review. Sadhana **38**(3), 461–476 (2013)
13. Winberg, C.: Teaching engineering/engineering teaching: interdisciplinary collaboration and the construction of academic identities. Teach. High. Educ. **13**(3), 353–367 (2008)

Chapter 10
Influence of Continuous Assessment Test Methodology on the Learning of Basic Mechanical Physics Knowledge

F. A. Berlanga, M. C. Vallejo, J. L. Borrego, and J. C. García-Prada

Abstract Two different methodologies are analysed for the implementation of continuous assessment tests in "*Physics I*" subject, which is taught for the introduction to mechanical physics, taught through the distance learning methodology. The results of the evaluation using the two methodologies in two different courses are analysed and the opinion of the students and the Mentor-Professors (MP) of the subject is analysed through the realization of surveys. One methodology (M1) is based on the asynchronous completion of a series of exercises related to the contents of the course. The other (M2) is based on the synchronous completion of online forms in which answers must be given to theoretical and practical questions about the contents of the course. The results show that there is a discordance between the methodology that gives the best results in terms of evaluation (M1) and the one preferred by the MP and students of the subject (M2).

Keywords Continuous assessment · Teaching methodology · Introductory mechanics

10.1 Introduction

The realization of Continuous Assessment Tests (CAT) is important for the learning of Physics Mechanics [3]. As it is a learning by assimilation and not very memoristic, the contemporizing of the development of the subject designed by the teaching team must be accompanied by the acquisition of competences by the student. For this it is necessary that the time dedicated to learning the subject is sufficient and coordinated with the work plan designed by the teaching team [4, 5]. If this does not occur, imbalances can be created in learning that end up creating knowledge gaps that compromise the progress of the student's understanding of the subject [4].

F. A. Berlanga (✉) · M. C. Vallejo · J. L. Borrego · J. C. García-Prada
Department of Mechanics, UNED, Madrid, Spain
e-mail: felixberlanga@ind.uned.es

J. C. García Prada et al. (eds.), *Trends in Educational Activity in the Field of Mechanism and Machine Theory (2018–2022)*, Mechanisms and Machine Science 128,
https://doi.org/10.1007/978-3-031-25730-8_10

To avoid this type of situation, it is positive to create a series of CAT that represent a milestone for the student to ensure that, on the one hand, he/she is making progress in the learning of the subject and, on the other hand, it represents a feedback from the Teaching Team Professors of the subject on his/her performance in the subject. The good design of the management of the CAT allows the student to redirect the direction of his learning before failing in one or more subjects [2].

In order for the learner to take advantage of CAT, they must be designed to be attractive and easy to take [6]. To this purpose, it is useful that these tests have an impact on the student's grade and that they are carried out in an accessible way. In this situation, the student will dedicate time to them and will be motivated for their preparation.

This work analyses the process of implementation of the CAT in the subject "*Physics I*" taught in all the degrees of the School of Industrial Engineering of the UNED. The impact of two methodologies is studied, a traditional methodology of non-synchronous resolution of proposed exercises and, on the other hand, a methodology that combines the response to questions related to the syllabus and the completion of exercises through web forms. On the one hand, the students' performance is analysed, and on the other hand, their opinion on the development of the process.

The objective of this study is to determine the goodness and shortcomings of each of the methods analysed and the possible improvements that can be implemented on them based on the data collected both in the performance of the students and in their opinion.

10.2 Methodology

10.2.1 Pedagogical Context

The subject under study is "*Physics I*", in which basic knowledge of Mechanical Engineering is taught so that students can face the advanced learning of Mechanics in higher courses, it is therefore a subject of vital importance for students to acquire the concepts that will allow them to face future learning without problems.

The teachings imparted in the subject are organized in four different modules: Kinematics, Dynamics, Fluid Mechanics and Gravitation and Thermodynamics. The module structure helps to organize the student's time and learning.

The subject is taught in UNED degrees, which presents a distance learning methodology. This methodology has been implemented for 50 years and has been improved over time. However, it is not exempt from various ills that afflict such education, such as high dropout rates by students or the low examination rate among enrolled students. These ills are usually attributed to the fact that distance students tend to relate differently to studies they are pursuing than the ones that are educated using on-site studies [1].

One of the means available to the university to alleviate the perceived distance between students and faculty is the existence of two types of student-facing faculty. At the university headquarters, several professors make up the Teaching Team for the subject, and they are in charge of designing the teaching–learning process for the student and evaluating their performance. On the other hand, the university has a network of Mentor-Professors (MP), who tutor students within Spain and some European capitals, and all UNED students around the world virtually. The involvement of the MP in the implementation plan of the CAT is considered vital for its success.

10.2.2 Description of the Methodologies Used to Carry Out the Continuous Assessment Tests

The implementation of the CAT for the subject "Physics I" has experienced two phases considering different methodologies. In both cases a CAT is proposed for each module in which the contents of the subject are divided, being in total four the number of CATs to be performed.

The initial methodology (*M1*) is based exclusively on the realization by the student of exercises on the subject in a non-synchronous way. A series of exercises is proposed that each student solves and returns to his MP before a deadline.

The second methodology implemented (*M2*) is based on the completion of a series of theoretical-practical exercises by the student in a synchronous way through a web form.

In both cases, the MP of the course oversees correcting and giving feedback to the student according to his performance. The grade obtained by the student is sent to the Teaching Team of the subject that evaluates it.

10.2.3 Analysed Parameters

The study analyses two types of results, those obtained through student evaluations and those obtained through student and MP surveys.

The values analysed in relation to student assessment are as follows:

- Success rate (*Sr*), understood as the proportion of students who pass the subject (*ast*) among those enrolled (*amt*).

$$St = ast/amt \tag{10.1}$$

- Evaluation rate (*Evt*), understood as the proportion of students who are fully evaluated (*aet*) of the subject out of those enrolled.

$$Evt = aet/amt \tag{10.2}$$

- Average grade (*Ct*) of the students who pass the course.

The comparison between students who complete the CATs (*p*) and those who do not (*np*) is also evaluated. In this case considering the relative success rate (*Sr*) and the average of the grades between students who pass the course having passed the CATs (*Srp*) and those who pass the course without having passed the CATs (*Srnp*).

$$Srp = asp/Evp;\ Srnp = asnp/Evnp \tag{10.3}$$

The results presented compare two different courses in which different CAT models were used. *M1* was used in the 2020/2021 academic year and *M2* was used in the 2021/2022 academic year.

In order to analyse the opinion of the students in relation to the two types of CAT, a survey is made to the students taking the course in the academic year 2021/2022 to analyse their opinion, and an analysis is also made on the MP of the course. Questions are asked to analyse the following factors:

- Opinion on the CAT implantation.
- Opinion on the weight of the grade in the final grade of the course.
- Opinion on different types of CAT models.

The surveys are carried out at the end of the 2021/2022 academic year among all of the subject's Tutor Professors and on a sample of students enrolled in the course during the 2021/2022 academic year.

10.3 Results

10.3.1 Student Performance Considering the Two Continuous Assessment Test (CAT) Methodologies

A general comparison of the results obtained through the evaluation of the students is shown in Fig. 10.1.

The results in Fig. 10.1 show that both the success rate and the evaluation and average grades for students who pass the course are better when the *M1* method is used for evaluation. The greatest differential between the three results presented is in the success rate of the students. However, this value is influenced by the evaluation rate, since only students who are evaluated can pass the course.

In terms of the grades obtained, there is a clear advantage for the *M1* method.

Figure 10.2 shows the comparison between students who took the continuous assessment tests and those who did not in their evaluation results

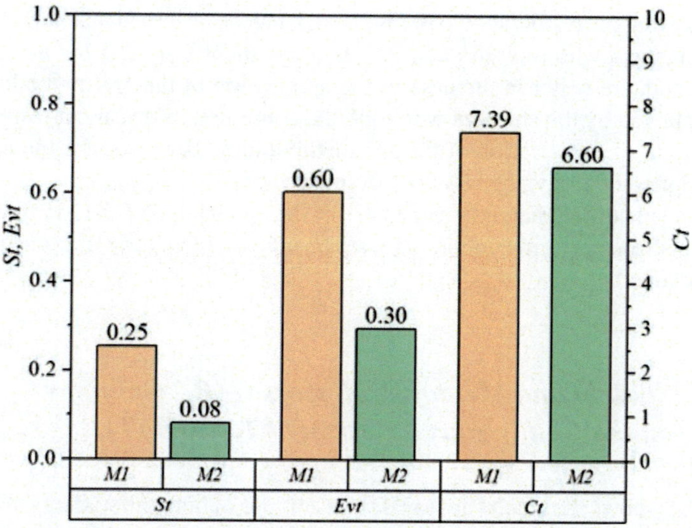

Fig. 10.1 Global comparison of evaluation indicators. *St*, success rate; *Evt*, evaluation rate; and *Ct*, average of grades among students who pass the course

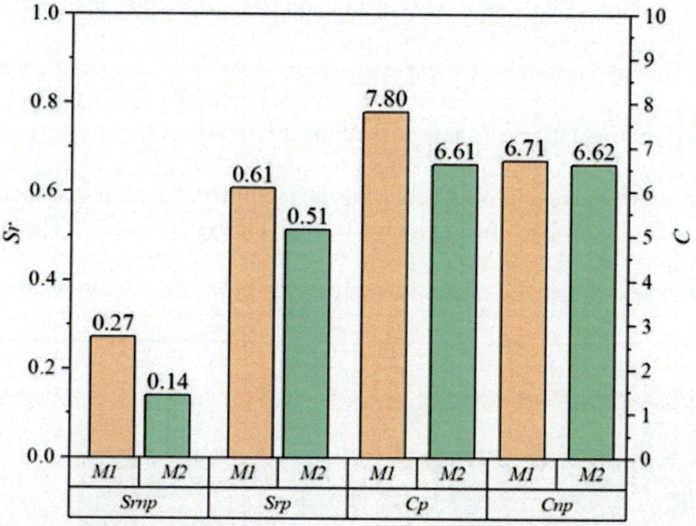

Fig. 10.2 Comparison of evaluation results between students who took the continuous evaluation tests and those who did not. *Sr* represents the relative success rate, and *C* represents the average grades of the students. The subscripts *p* and *np* represent the students who took the continuous assessment tests and those who did not

Considering the results shown in Fig. 10.2, it can be observed that both the success rate for students who took the continuous assessment tests and for those who did not take them are better in the case of the application of the *M1* methodology. The grades obtained by the students who took the continuous assessment tests are better in the case of the application of the *M1* methodology. However, for the application of the *M2* methodology, there is no difference.

It can be identified that only in the case of the application of the M1 methodology for students who perform continuous assessment tests there is an improvement in the results obtained.

10.3.2 Student and Mentor Professors (MP)' Opinions on the Continuous Assessment Tests (CAT)

Figure 10.3 shows the results obtained from surveys of both the students of the course and the tutors of the course. The results are based on the survey of 45 students and 10 tutors of the course.

Figure 10.3a shows the results collected on the opinion of both students and *MP* on the application of the continuous evaluation tests. The opinion is evaluated from 0 to 5, with 0 being the worst opinion and 5 the best.

Two different questions are presented, one to identify the opinion they have of the use of continuous assessment tests for the evaluation of the subject (*Form. Opinion*) and the weight that these tests have in the final evaluation of the subject (*Calif Weight Opinion*).

According to the results obtained, it can be identified that both students and tutors have a good opinion about the implementation of the system of CAT. The students are

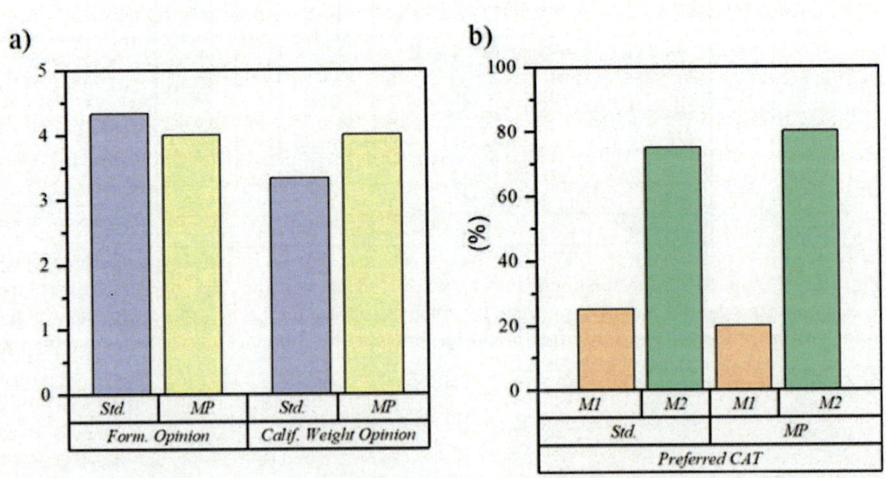

Fig. 10.3 Results from surveys. Where *Std.*, represents students and *MP*, mentor-professors

more enthusiastic about the implementation of the system, while the tutor Professors value more its inclusion in the evaluation.

Figure 3b shows the opinion on the preferred methodology for carrying out the continuous assessment tests. Both groups, students, and professors-tutors, prefer to carry out the continuous assessment tests using the *M2* method.

10.4 Conclusions

Based on the results obtained, the following conclusions can be identified:

- The results obtained from the general evaluation of the student body point to the fact that the use of the *M1* methodology, based on the resolution of non-synchronous exercises, improves all the performance indicators of the students evaluated.
- The comparison between students who take the Continuous Assessment Tests (CAT) and those who do not, using the two methodologies, indicates that the success rate and average grades increase when using the *M1* methodology.
- The opinion surveys both to the Mentor-Professors (MP) and to the students of the subject indicate that the preferred methodology for taking the CAT is the *M2* methodology, based on the completion of synchronous online forms.

The results obtained in this work indicate that the results obtained by the students in the evaluation improve using the CAT *M1* methodology, however the generalized perception is that the *M2* methodology is the most indicated for the realization of the mentioned tests.

References

1. Cano, E.V., Garca, M.S.: Analysis of risks in a learning management system: a case study in the Spanish National University of Distance Education (UNED). J New Approach. Educ Res NAER J **4**, 62–68 (2015)
2. Catalán Cantero, C., Gallardo, J., Blesa, A.: Análisis comparado de normativas de evaluación: cuestiones abiertas. Asociación de Enseñantes Universitarios de la Informática (AENUI) (2017)
3. Ganiel, U., Hofstein, A.: Objective and continuous assessment of student performance in the physics laboratory (1981)
4. Hernández, R.: Does continuous assessment in higher education support student learning? High. Educ. **64**, 489–502 (2012). https://doi.org/10.1007/s10734-012-9506-7
5. Islam, K., Ahmadi, P., Yousaf, S.: Assessment formats and student learning performance: what is the relation? (2017)
6. Santos, J.-M., Ortiz, E., Marín, S.: Variation indexes of marks due to continuous assessment. Empirical approach at university/Índices de variación de la nota debidos a la evaluación continua. Contrastación empírica en la enseñanza universitaria. Cult. Educ. **30**, 491–527 (2018). https://doi.org/10.1080/11356405.2018.1488422

Chapter 11
Evaluation of Audiovisual Guides for Laboratory Classes in Hydraulic Machinery Courses of Distance Learning Engineering Programs

Claudio Zanzi, Pablo Gómez, Félix Berlanga, and Julio Hernández

Abstract In this work, audiovisual guides for laboratory classes on hydraulic machinery in distance learning engineering programs are developed and evaluated. The aim of these guides, which can be viewed directly within the virtual classroom via computer, tablet or mobile phone, is to provide students with an accurate knowledge of the tasks they will have to perform and of the equipment they will handle during the laboratory classes, in order to make the best possible use of the available time during the lab sessions. A key aspect in the development of the guides was to achieve a final format with a relative small size and, at the same time, a reasonable good image and sound quality. In addition, the guide is intended to provide the learner with the highest possible degree of self-sufficiency. The efficiency of the guides is assessed by means of the results of questionnaires answered by students before starting the laboratory classes and by analyzing the development of the lab sessions.

Keywords Distance learning engineering · Audiovisual guides · Laboratory practices · Hydraulic machinery

11.1 Introduction

Throughout history the laboratory education has been identified as a key part in the scientific and engineering education [1–3]. Engineering is a practicing profession and laboratory classes usually represent the first contact of students with materials and technical equipment that will be part of the future profession. Through interaction with materials and handling of equipment, students are able to construct their knowledge of physical phenomena and scientific concepts [4]. The role of laboratory

C. Zanzi (✉) · P. Gómez · F. Berlanga · J. Hernández
Departamento de Mecánica, ETSII, Universidad Nacional de Educación a Distancia (UNED), E-28040 Madrid, Spain
e-mail: czanzi@ind.uned.es

J. C. García Prada et al. (eds.), *Trends in Educational Activity in the Field of Mechanism and Machine Theory (2018–2022)*, Mechanisms and Machine Science 128,
https://doi.org/10.1007/978-3-031-25730-8_11

classes is to develop students' learning and ability to apply the theory into practice and help them to better assimilate theoretical concepts [5].

One of the main problems of distance learning in engineering refers to limitations related to practical and laboratory work [6]. When undergraduate distance learning programs in engineering started to develop, different solutions began to be implemented to give students the necessary laboratory experience [2, 7], such as conducting laboratory classes at different institutions (e.g., a local educational institution like a college or a high school with the necessary facilities); programming concentrated courses in physical laboratories on campus, where conventional laboratory facilities can be used; providing students who live far from campus with lab kits so that they can perform lab experiments at home; and, more recently, with the development of the Internet, the use of physical laboratories with remote access or virtual laboratories that simulate physical facilities using simulation software. Each solution has its own advantages and disadvantages; for example, the first does not require much travel from home, although it requires specialized teaching staff only available in courses of the first years of higher education. The second and third approaches involve additional costs due to the travel to the city where the campus is located or the purchase of lab kits, respectively. On the other hand, online labs offer great advantages in engineering laboratory learning and become an alternative to physical labs; however, they do not always allow to acquire the same competences, and in many cases they are used as a complement [8, 9]. An extensive comparative analysis between online and physical labs can be found in [6].

One of the features of the teaching of hydraulic machinery courses, common to most of other scientific-technical courses, is the need to carry out laboratory classes, allowing students to handle the basic instrumentation used in hydraulic machinery, to observe the operation of different hydraulic turbomachines under several operating conditions, to experimentally verify the basic principles studied in the theoretical part of the course, and to determine their characteristic curves. These classes are carried out in engineering schools with on-site teaching during several face-to-face laboratory sessions. In the case of UNED, which combines online teaching and on-site support, the laboratory equipment necessary to perform this type of face-to-face sessions is only available in the laboratory of the central campus. In order to limit the number of times that students need to travel to the central campus from different cities in the country (or, in certain cases, from other countries), it was decided to limit the laboratory classes for each course to a maximum of four sessions of four hours, thus making it necessary to take maximum advantage of the time during the labs. To this purpose, students are currently sent a lab class script in advance, allowing them to study it and to review the required theoretical contents to be used during the lab classes. However, these scripts do not give students a sufficiently accurate idea of the equipment to be handled and of the experimental procedures to be followed, and it is in this context that it has been decided to carry out this work, whose main goal is to achieve the best possible use of time during the laboratory classes.

For this purpose, audiovisual guides have been developed for laboratory classes, and have been made available to students through the virtual classroom via a multimedia file, which can be viewed by computer, tablet or mobile phone. One of the key aspects has been to keep the size of the files reasonably small, which makes a video recording unfeasible and requires considering different alternatives. The audiovisual guides describe the laboratory equipment, the procedure to be followed and the results to be obtained. They allow viewing the equipment and describe its configuration, the control panel, the readings of the measuring instruments under various operating conditions, and the start-up procedure.

In a first step, we elaborated a preliminary guide on the testing of a centrifugal pump, and evaluated its utility and efficiency with a questionnaire that was answered by the students at the beginning of each laboratory session. The questionnaire focused on several aspects, such as the time spent by the students in viewing the guide, their degree of assimilation of the contents achieved, and their evaluation of the guide. Then, based on the results obtained, the first guide was improved and a second guide was prepared on the testing of a Pelton turbine, in an attempt to correct the problems detected in the first guide and to achieve a greater degree of self-sufficiency of the explanations of the guide.

11.2 Development of the Audiovisual Guides

11.2.1 Guides General Description

The final result of this work consists of two audiovisual guides in AVI format, of approximately seven and ten minutes of duration, respectively, that the students will have to watch before attending the laboratory classes. The guides describe and explain the experimental procedures of two laboratory tests common to several courses belonging to the Fluid Mechanics Area of the Department of Mechanics of UNED. The main goal of these tests is to obtain the characteristic curves and performance results of a centrifugal pump and a Pelton turbine installed on a hydraulic bench. To achieve this objective, it is necessary to analyze the data taken from a set of readings of different measuring equipments, for several operating conditions of the pump and turbine.

Both guides show a sequence of slides in which photographs and images are described and commented by a voice-over. To facilitate the understanding of the guide, indicators are included on the images that appear as the voice-over explains the content of each slide. The guides begin by stating the objectives of the test, then describe the main components of the hydraulic bench, and finally show the steps that have to be followed to operate the control elements of the bench, read the data, analyze them and obtain the characteristic curves and performance results of the centrifugal pump and the Pelton turbine. In the theoretical study, students use textbooks such as those of Refs. [10–13].

11.2.2 Guides Elaboration Procedure

The general steps followed for the elaboration of the guides are described below.

11.2.2.1 Photographs

Pictures of the most relevant elements of the hydraulic bench were taken and edited using the Paint Shop Pro image editor to obtain a better quality.

11.2.2.2 Power Point Presentation

A Power Point presentation was created using the edited pictures; one or more images are shown on each slide of the presentation, along with indicators to specify which elements are described.

11.2.2.3 Audio

A script was written to describe the test and the elements appearing in the images and used to record and edit a voice-over using the WavePad Sound Editor software. This program allows the resulting audio file to be recorded in different formats, including WAV and MP3.

11.2.2.4 Video Editing

Once the Power Point presentations and audio files were ready, we proceeded to associate each slide with the corresponding audio file, so that the presentations would have an automatic transition between slides and in synchrony with the explanations of the audio files. The fact that Power Point presentations can only embed audio files in WAV format would make the final size of the guide too large and impractical for downloading over the Internet. So it was decided to record the sound in MP3 format, which takes up less space than the equivalent WAV format, and mount it together with the presentations using the E.M. Power Point Video Converter program, which allows you to convert the presentation together with the audio to various video formats choosing from a multitude of recording options. We opted for a result that offered reasonably good image and audio quality, keeping the size of the output file as low as possible. The final result was, for each guide, an AVI video file with the audio in MP3 format, with a resolution of 640 × 480 pixels. The duration of the videos is seven minutes for the guide on the pump test, and ten minutes for the guide on the turbine test. This format is also very convenient for the student if he/she wants to go back or forward in the visualization. Figures 11.1, 11.2, 11.3 and 11.4 show a few

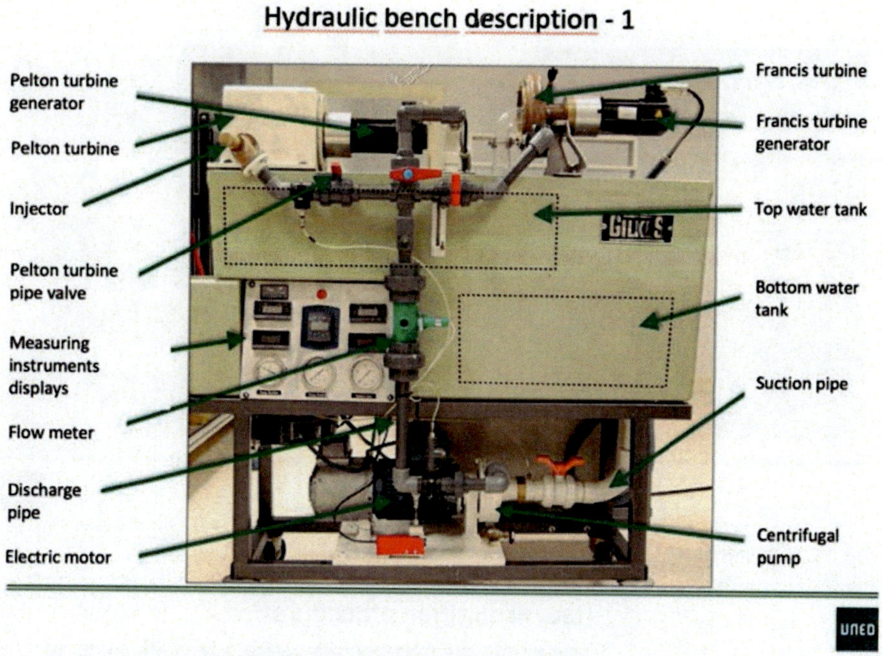

Fig. 11.1 Description of the hydraulic bench

representative Power Point slides from both guides that were used to produce the final videos. Below each slide (except for the cover slide) we included the transcription of the corresponding audio.

11.3 Evaluation of the Preliminary Guide on the Pump Test

As mentioned above, to evaluate the utility and efficiency of the first preliminary audiovisual guide on the centrifugal pump test, a questionnaire was given to students at the beginning of each lab session. Students who, for any reason, had not watched the video yet, were let to do so on a laptop in the laboratory to get a result from the questionnaire based on the answers of as many students as possible. The questionnaire was made anonymous so that students could feel able to answer more freely. The questions in the questionnaire focused on the following three aspects:

- Time spent by the students viewing the guide.
- Degree of assimilation achieved of the contents explained.
- Evaluation of the guide by the students.

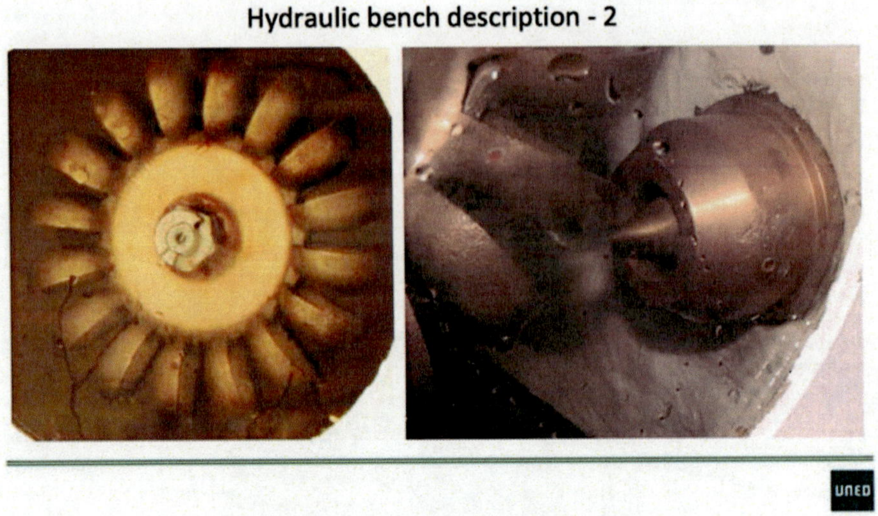

Fig. 11.2 Pelton turbine. Wheel (left image) and noozle (right image)

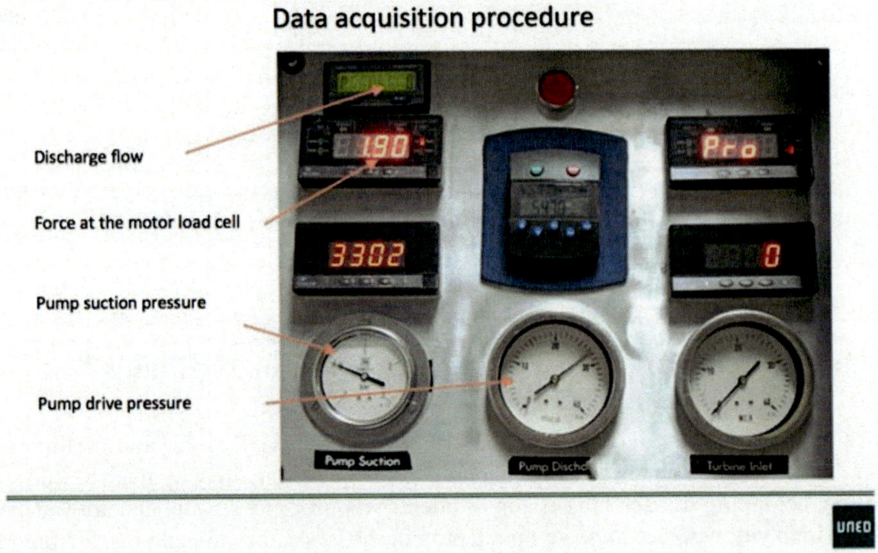

Fig. 11.3 Data acquisition procedure

The degree of assimilation of the contents explained in the guide was assessed by a test of four questions, included at the end of the questionnaire, related to the main objectives of the test, the components of the hydraulic bench and the experimental procedure to be followed during the test.

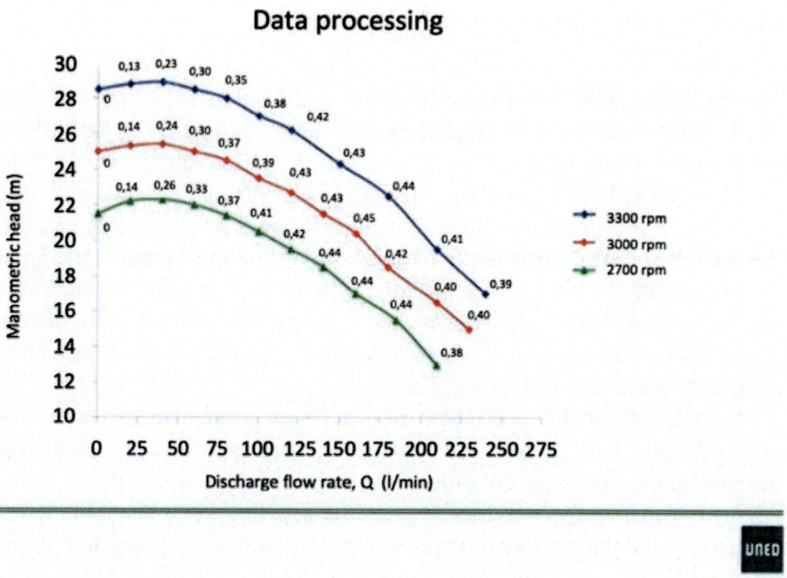

Fig. 11.4 Data processing. Characteristic curves and performance results

11.3.1 Results of the Questionnaire on the Preliminary Guide

From the analysis of the results obtained from the questionnaire, the following conclusions can be drawn:

- The image quality appears to be good.
- The sound quality has a poorer rating. This may be due to the fact that not all students have equipment with good audio components.
- The time spent studying the guide is reasonably small (less than 30 min in all cases).
- 54% of the students watched the video twice. Considering that the pump and turbine video tutorials have a duration of 7 and 10 min respectively, this means a dedication of 34 min.
- More than 40% of the students required additional explanations. It is interesting to note that these are precisely the students who watched the video in the laboratory, which seems to indicate that, when watched at home, comprehension is better.
- In the test of knowledge of the contents explained in the guide all students obtained high scores, which seems to indicate that the video was quite clear and that they followed it attentively. This would explain that only 4% of the students failed the test and that the percentage that scored above 8 out of 10 was 92%.

11.4 Pelton Turbine Guide and Survey

The experience gained from the first guide on the pump test allowed to improve it and to take it into account in the elaboration of the second guide on the turbine test. The main objective of this second guide was to make its explanations as self-sufficient as possible, not only in the description of the elements and the operation of the bench, but above all in the experimental procedures to be followed by students, thus reducing the time required to perform the test and improving the students' independence.

To evaluate the degree of utility and efficiency of the guide, the development of the lab sessions was analyzed, observing the performance of three different groups of students who were given different degrees of on-site explanations about the procedures to be carried out during the test.

Another aspect that was taken into account is the fact that the turbine test is more complex than the pump test, as it requires the use of a second control panel to regulate the operating conditions of the generator coupled to the turbine and the energy dissipation system. In this case it is necessary to act simultaneously on two controllers and the number of explanations to describe the procedure is considerably greater.

11.4.1 Results Obtained with the Pelton Turbine Guide

The three groups of students mentioned above were labelled as groups *A*, *B* and *C*. Before starting the lab session, it was verified that all students had previously watched at home the audiovisual guide on the Pelton turbine testing. Then, the test was explained to students by the teaching staff. The groups in each session received a different level of explanation: group *A* was given a full explanation of the test, group *B* a partial explanation and group *C* no explanation at all. Table 11.1 shows the times and type of queries raised by each of the groups during the lab session. The average viewing time corresponds to an estimate of the average time spent by the students viewing the audiovisual guide for this test, which lasted approximately 10 min.

The monitoring of the development of the laboratory sessions showed that the audiovisual guide for the Pelton turbine test allows the students to perform during the lab session in a self-sufficient way.

The type of questions asked by students of the three groups were as follows:

- Group *A*: Few questions were raised, all of them related to aspects of the analysis of the experimental results. No questions related to experimental procedure were asked. There were also questions related to the theoretical foundations of the course, which were not covered either in the on-site explanations or in the guide.
- Group *B*: A question related to the procedure was asked (specifically, about the number of samples to be taken for each case), and also a number of questions related to the elaboration of the graphs and the theoretical foundations.

Table 11.1 Results obtained for the Pelton turbine test guide

Groups	A	B	C
Number of students	11 (3 groups)	3 (1 group)	5 (1 group)
Explanation	Complete	Partial	None
Time dedicated to explanation (min)	20	10	0
Average time of visualization (min)	22	22	22
Type of questions			
Bench operation	–	–	X
Experimental procedure	–	X	–
Analysis of results	X	X	X
Theoretical foundations	X	X	X

– Group *C*: Several doubts were raised about the operation of the bench, revealing shortcomings in the explanations given in the audiovisual guides (specifically, it was found that the video did not show how to connect the pressure gauge at the turbine inlet). This procedure explanation had been described in the on-site explanations, both for groups *A* and *B*. Although some doubts were raised about the results analysis and theoretical foundations, similar to those raised by students in groups *A* and *B*, no questions about the procedure were asked. One of the most relevant aspects found is that the students were able to perform the entire data acquisition part with independence, rarely requiring on-site explanations, in the same way as the students in the other two groups.

It is interesting to note that students in group *B* asked questions about the experimental procedure, while those in group *C* did not. One possible explanation is that having two sources of information (audiovisual guide and on-site explanation) may have introduced some confusion, especially if the second explanation is not complete or is provided too quickly.

11.5 Conclusions

In this work, two audiovisual laboratory guides were developed for distance learning engineering students. The main purpose was to provide students with an accurate knowledge of the equipment to be handled and the experimental procedure to be performed during the laboratory classes of a hydraulic machinery course, in order to make the best possible use of the time available during the lab sessions.

The results obtained with a first preliminary guide, focused on the testing of a centrifugal pump, indicate that the students understood reasonably well the theoretical contents explained in the guide, which has led to a better preparation of the students for the classes, with the consequent improvement in their performance and, therefore, in the assimilation of the course contents. However, during the realization of the tests, a lack of information of the students on some aspects of the experimental procedure to be carried out was detected.

In the elaboration of a second guide, on the testing of a Pelton turbine, an attempt was made to reduce the limitations detected in the first preliminary guide. To assess the degree of self-sufficiency of the audiovisual guide, the development of the lab sessions was analyzed, confirming that with the explanations of the guide the students were able to carry out without help all the part of the test related to data collection and instrumentation handling. It has also been found that, although a complete explanation of the test is given on site, it is necessary to repeat the explanations related to the elaboration of the final results.

In general, the audiovisual guides allowed the students to achieve a better preparation to carry out the laboratory tests, with the consequent improvement in their performance during the lab sessions. These conclusions have been confirmed by reviewing the results reports presented by the students and comparing them with those of previous years, as well as by the degree of student involvement observed during the lab classes.

References

1. Gibbins, L., Perkin, G.: Laboratories for the 21st Century in STEM Higher Education. Loughborough University, Centre for Engineering and Design Education (2013)
2. Feisel, L.D., Rosa, A.J.: The role of the laboratory in undergraduate engineering education. J. Eng. Educ. **94**(1), 121–130 (2005)
3. Edward, N.S.: The role of laboratory work in engineering education: student and staff perceptions. Int. J. Electr. Eng. Educ. **39**(1), 11–19 (2002)
4. Robin, K.: Research on science laboratory activities: in pursuit of better questions and answers to improve learning. Sch. Sci. Math. **90**(5), 403–418 (1990)
5. Goodwin-Jones, J., Nortcliffe, A., Vernon-Parry, K.: What does good engineering laboratory pedagogy look like? Engineering education on top of the world: industry university cooperation. In: 44th SEFI Conference Proceedings. SEFI (2016)
6. Stefanovic, M.: The objectives, architectures and effects of distance learning laboratories for industrial engineering education. Comput. Educ. **69**, 250–262 (2013). https://doi.org/10.1016/j.compedu.2013.07.011
7. Bengiamin, N.Y., Johnson, A., Zidon, M., Moen, D., Ludlow, D.K.: The development of an undergraduate distance learning engineering degree for industry—a university/industry collaboration. J. Eng. Educ. **87**(3), 277–282 (1998)
8. Foreman, C., Hilditch, M., Rockliff, N., Clarke, H.: A comparison of student perceptions of physical and virtual engineering laboratory classes. In: Enhancing student-centered teaching in higher education, pp. 151–167 (2020). https://doi.org/10.1007/978-3-030-35396-4
9. Balamuralithara, B., Woods, P.C.: Virtual laboratories in engineering education: the simulation lab and remote lab. Comput. Appl. Eng. Educ. **17**, 108–118 (2009)

10. Hernández Krahe, J.M.: Mecánica de fluidos y máquinas hidráulicas, Unidades Didácticas 5ª y 6ª. UNED (1976) (last available edition: 1995)
11. Mataix, C.: Turbomáquinas hidráulicas. Ed. ICAI (1975)
12. Crespo, A.: Mecánica de fluidos. Thomson (2006)
13. Hernández, J., Gómez, P., Zanzi, C.: Máquinas hidráulicas. Problemas y soluciones, UNED (2016)

Chapter 12
Learning How to Design and Manufacture by Applying Hot Wire Cutting to the Fabrication of a Car Spoiler

A. M. Gómez-Amador, S. Sanz, D. García-Pozuelo, and V. Díaz

Abstract In order to introduce the students of the Mechanical Engineering Degree in the design and manufacture of a mechanical element, it is proposed the development of a practice to obtain the spoiler of an automobile vehicle. The technique to be used will be hot wire cutting; this technique, although known by the students, has not been used during their studies or practices of Degree. Hot wire cutting is a manufacturing technique that basically consists of heating a nichrome wire by electrical resistance. The profiles will be manufactured and a metallic arch will be used, where the nichrome wire will be heated to melt blocks of extruded polystyrene foam following the contour of these aerodynamic profiles to finally extract the core. The aim is for the student to put into practice the technical knowledge acquired during his studies and to develop personal skills that can help them in their professional life. The best way for a student to be stimulated and to have an adequate learning process is to complement the theoretical sessions with practical sessions.

Keywords Mechanical design · Aerodynamics · Hot wire cutting

12.1 Introduction

The purpose of this practice is to introduce engineering students to the design and manufacture of mechanical elements with complex geometries, such as spoilers and wings, by means of hot wire cutting.

A. M. Gómez-Amador (✉) · S. Sanz · D. García-Pozuelo · V. Díaz
Department of Mechanical Engineering, University Carlos III de Madrid, 28911 Madrid, Spain
e-mail: amgomez@ing.uc3m.es

S. Sanz
e-mail: ssanz@ing.uc3m.es

D. García-Pozuelo
e-mail: dgramos@ing.uc3m.es

V. Díaz
e-mail: vdiaz@ing.uc3m.es

J. C. García Prada et al. (eds.), *Trends in Educational Activity in the Field of Mechanism and Machine Theory (2018–2022)*, Mechanisms and Machine Science 128,
https://doi.org/10.1007/978-3-031-25730-8_12

Hot wire cutting is a manufacturing technique that basically consists of heating a nichrome wire (80% nickel, 20% chromium) by electrical resistance. This material has a high resistivity and a high melting point ($100 * 10^{(-8)}$ Ωm). It is used to create resistors in a multitude of applications (dryers, toasters, electric ovens, etc.).

In this practice, as will be detailed in the following sections, some aerodynamic profiles will be made in wood and a metallic arch with nichrome wire that when heated will melt the XPS (extruded polystyrene) foam blocks following the contour of these profiles; once this process is finished, the cores will be extracted. For this purpose, a power supply will be used to make pass the current through the wire and heat it up.

In these practical sessions, active learning techniques will be applied, i.e., the principles cited by Confucius in the fifth century B.C. [1]: "I was told and I forgot it; I saw it and I understood it; I did it and I learned it".

The aim is to stimulate the student's creativity, apply technical concepts to practical issues, enhance teamwork skills and manage emotions in a group. The selection of this topic as the reason for the practice is a response to the interest shown by the students regarding the mechanics of an automobile and the participation in competitions carried out by the Universidad Carlos III de Madrid (specifically the competitions in which the students integrated in the Formula Student Group participate (see Fig. 12.1).

The problem is first approached theoretically, the equations and concepts that govern the passage of a solid through a fluid, air, are presented. And then the templates of the ailerons and the arch are made to proceed to the cutting of the polystyrene and thus obtain the core.

Fig. 12.1 Vehicle designed by the Formula Student Team of the University Carlos III of Madrid, MAD Formula Team

12.2 Theoretical Concepts: Fundamental Parameters

Wooden horse-drawn carriages began to be used in the 16th century. Two centuries later, in order to improve the stability, aesthetics and durability of the vehicle, materials such as steel and aluminum began to be used. The continuous evolution in this respect has led to great advances in the design and functionality of the bodywork. Today, beyond aesthetics, the aim is for the body to be aerodynamic, light and comfortable, and, of course, without forgetting its functionality.

Ferrari [2] indicated in his book that aerodynamics was only for losers who did not know how to make engines and when a customer bought a Ferrari, he was paying for the engine, the rest was given for free. Today this thinking has evolved: to the improvements incorporated in the engine (power, consumption, etc.) we must add the advances in body design and the improvement in aerodynamic devices.

To design a car spoiler it is necessary to know the main variables with which the different airfoil families are determined and classified. Each of these families defines a certain characteristic geometry and its application.

Figure 12.2 shows a schematic of a car spoiler that generates lift. It could therefore be used in applications such as aircraft wings, or in racing car airfoils if its curvature is inverted.

Analyzing each of these variables:

Leading edge: This is the first point of the airfoil to come into contact with the incident wind at relatively normal angles. At this point, generally, there are areas of high pressure and low or no velocities. It will therefore be a stagnation point (P = Pmax, V = 0).

Trailing edge: This is the point furthest from the leading edge. Here, when we have a flow totally adhered to the airfoil, the flow that circulated on the lower surface will join again with the flow on the upper surface. Sometimes it will be interesting to delay this junction to achieve a higher force (or downforce, as in the case of the Gurney flap).

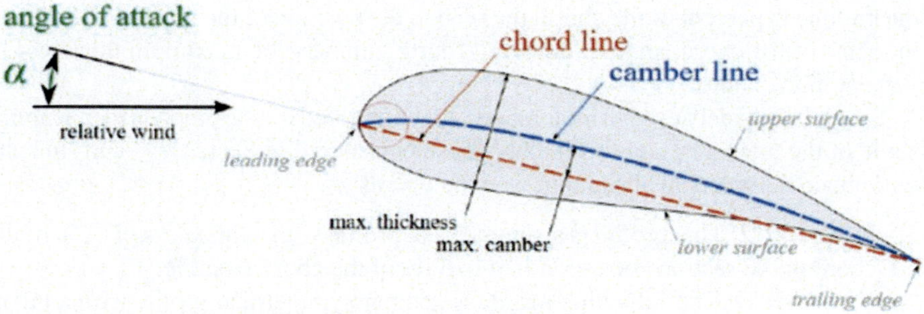

Fig. 12.2 Characteristic parameters in a car spoiler [3]

Chord: The chord is the length of the straight line joining the start and end of the profile, (or leading and trailing edge). The chord is used to characterize the flow by Reynolds number.

Mean line of curvature: It represents the succession of points located in the central coordinate with respect to the thickness of the profile. A profile will only be symmetrical if its curvature line and its chord coincide.

Angle of attack: Although this is not a car spoiler variable, but an airflow variable, it is a fundamental parameter. It is the angle that the incident wind forms with respect to the chord of the airfoil.

This type of profiles are normed, which allows a parametric design of them. This means that, by assigning numerical values to the mentioned variables, its point-to-point graph can be obtained. That is, by giving values to the different parameters of the mathematical function, the coordinates of all its points are obtained (an example of this are the NACA (National Advisory Committee for Aeronautics) airfoils). With these coordinates, it is possible to work with the profile in computer-aided design (CAD) and 3D printing model manufacturing programs.

12.2.1 Geometries Selected for Airfoil Fabrication

Since the practice is carried out with groups of 4 or 5 students, several NACA profiles were selected, the predecessor organization of the current NASA. These profiles were designed using high Reynolds numbers (dimensionless parameter that gives the relationship between inertia forces and those due to viscosity). NACA airfoils are classified and coded in a catalog; they are identified by the words NACA and a set of digits, in a certain order, with which their geometrical dimensions are determined. The aerodynamic characteristics of these profiles are given by the corresponding curves.

Thus, for example, for four-digit NACA profiles, the four numbers are based on the profile geometry [4]. Thus, the first number indicates the maximum ordinate of the midline in percent of the chord, the second the position of the maximum ordinate in tenths of the chord, and the third and fourth numbers the maximum thickness in percent of the chord.

Six airfoils widely used in motorsport will be selected for the practice to exemplify some of the main parameters that have been considered in Sect. 12.2. The students will obtain these airfoils from thin wooden boards:

1. NACA 0012: This profile is a symmetrical profile with a thickness of 12% of the chord in the section corresponding to 30% of the chord (see Fig. 12.3).
2. NACA E169: The following profile is another symmetrical profile with a thickness of 14.4% in the section at 26.5% of the chord. This profile is designed to operate at low Reynolds numbers, where the boundary layer is laminar and is resistant to disturbances that cause a transition to turbulent boundary

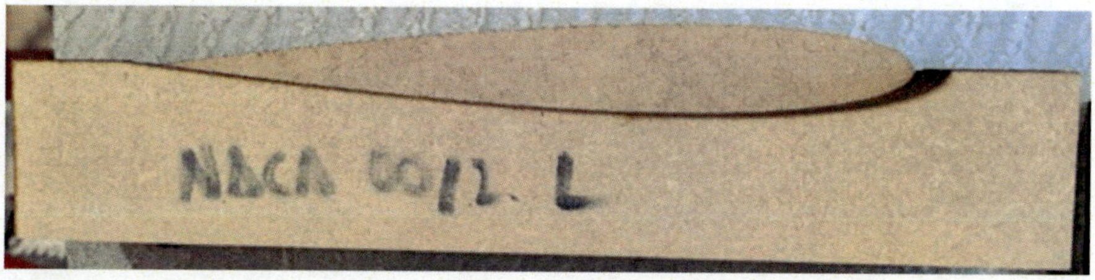

Fig. 12.3 NACA 0012. http://airfoiltools.com/airfoil/details?airfoil=n0012-il

layer, causing separation without subsequent re-entrainment due to large adverse pressure gradients (see Fig. 12.4).

3. NACA E423: The Eppler 423 airfoil is an asymmetric airfoil with a maximum camber of 9.5% at 41.4% of the chord and a maximum thickness of 12.5% at 23.7%. It is a profile designed to generate lift by optimizing efficiency in low Reynolds number conditions (see Fig. 12.5).
4. NACA FX 63-137: The following airfoil is an asymmetric airfoil with a thickness of 13.7% at 30.9% of the chord, with a maximum camber of 6% at 53.3% of the chord. It is a profile designed to generate aerodynamic loading at low Reynolds numbers with a smooth stall transition (see Fig. 12.6).
5. NACA FX 74-CL5-140: The FX 74-CL5-140 airfoil is an asymmetric airfoil with a thickness of 13.1–27.1% and a maximum camber of 9.7–41.6% of the chord. It is designed for low Reynolds numbers and, of all those present, is the airfoil that

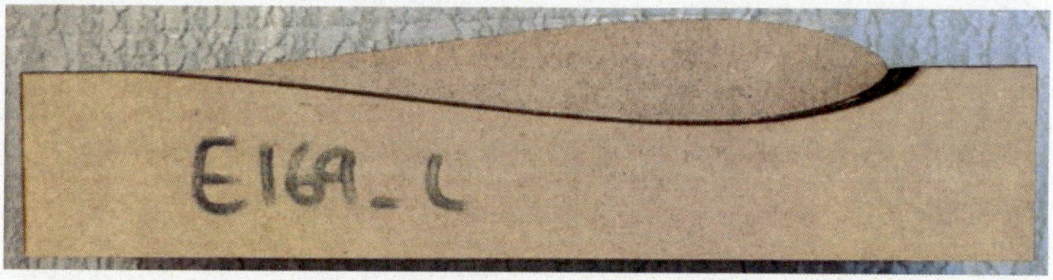

Fig. 12.4 E169. http://airfoiltools.com/airfoil/details?airfoil=e169-il

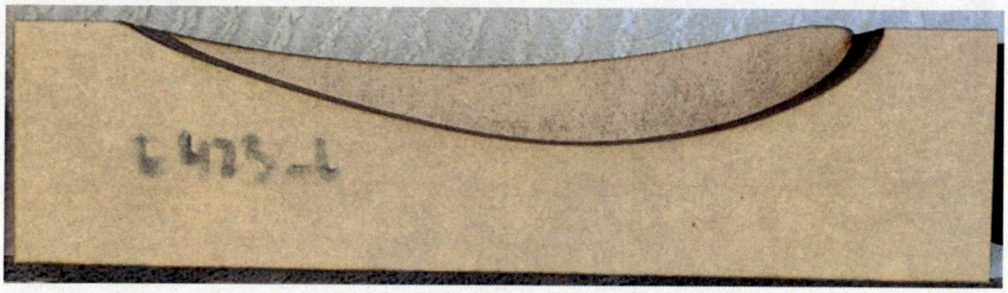

Fig. 12.5 E423. http://airfoiltools.com/airfoil/details?airfoil=e423-il

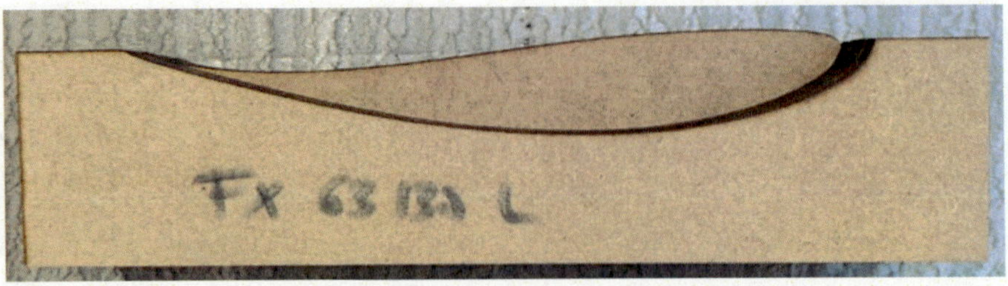

Fig. 12.6 FX 63-137. http://airfoiltools.com/airfoil/details?airfoil=fx63137-il

generates the most aerodynamic load in its performance range. However, it has a lower aerodynamic efficiency and is much more sensitive to lift loss for large angles of attack compared to the other proposed airfoils (see Fig. 12.7).

6. NACA LNV109A: The last airfoil is another asymmetric airfoil designed to maximize the lift coefficient with a perfectly adherent flow over the entire surface thanks to the concave recovery zone on the suction face of the airfoil, which allows smoothing the adverse pressure gradient in that region. It has a maximum thickness of 13% for 23.5% and a maximum camber of 6% for 31.5% of the chord (See Fig. 12.8).

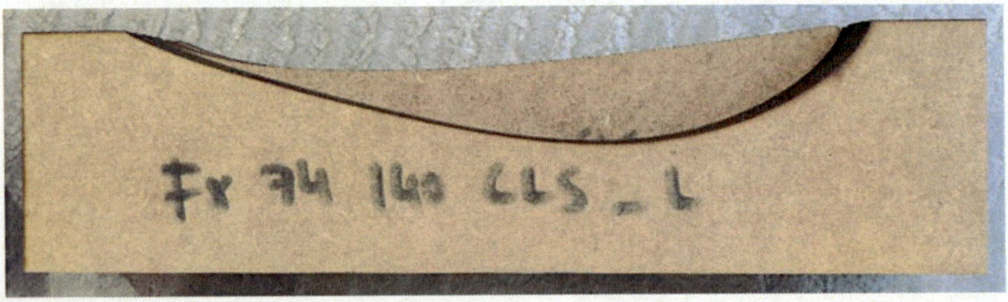

Fig. 12.7 74-CL5-140. http://airfoiltools.com/airfoil/details?airfoil=fx74modsm-il

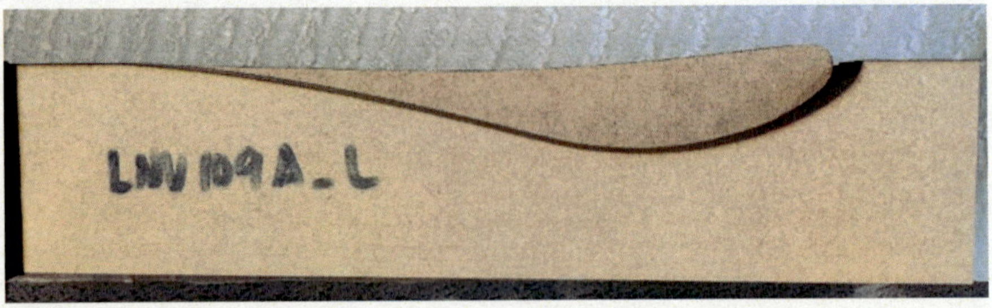

Fig. 12.8 LNV109A. http://airfoiltools.com/airfoil/details?airfoil=lnv109a-il

12.3 Cutting Arch Fabrication

The following materials and tools are required to make an arch for hot wire cutting:

- 1 spring.
- 1 3 mm DM wood board.
- 1 aluminum square profile of 30 mm of 1 m length.
- Screws (metric 4 or 6).
- Drill and drill bits (for wood and for aluminum) of the appropriate metric for the hardware.
- Fixed wrenches of the appropriate metric.
- Jig saw or hacksaw.

The fabrication of the arch will be done by following the steps below:

1. Cut with the jigsaw or saw four 'L' shaped wooden arms of the same length and a width of about 3–4 cm. The base of the 'L' should be approximately 10–15 cm long.
2. Drill the four arms at the same time in the following way: first, mark the two holes at the base of the 'L' so that there is a distance of 1–1.5 cm with the edges of the wood. Then, mark a hole in the middle of the arm that will be used for the spacer and another hole at the end of the arm for the thread. Finally, fix all the arms together with a pair of clamps and drill them all at the same time.

   The ends of the aluminum profile are drilled marking the position of the drilled holes of the base of the wooden arms.
3. The assembly is carried out according to the steps in the previous section. If no assembly screw or dowel screw is available, conventional screws can be used and the spring can be used to tension the wire without wrapping it around the screws. To do this, the appropriate length of wire should be cut so that it is just under the right length and the spring will allow some tension. Cut a length of cable about 15% longer than the distance from the spring tip to the dowel screw.
4. Place the washer on the other end of the spring, pass the cable through the outside of the pivot, pass it through the screw hole at the other end and wrap the cable around the screw (see Fig. 12.9).
5. Adjust the tension of the cable by turning the dowel screw so that it is slightly taut (a precise tension value is not necessary, as a reference, it should be slightly less than that of a guitar string) and then tighten the screw with a wing nut (see Fig. 12.10).
6. With the arch completely assembled, the rest of the components of the workstation are installed for the practical exercises, as shown in Fig. 12.11.

Fig. 12.9 Wire tension spring

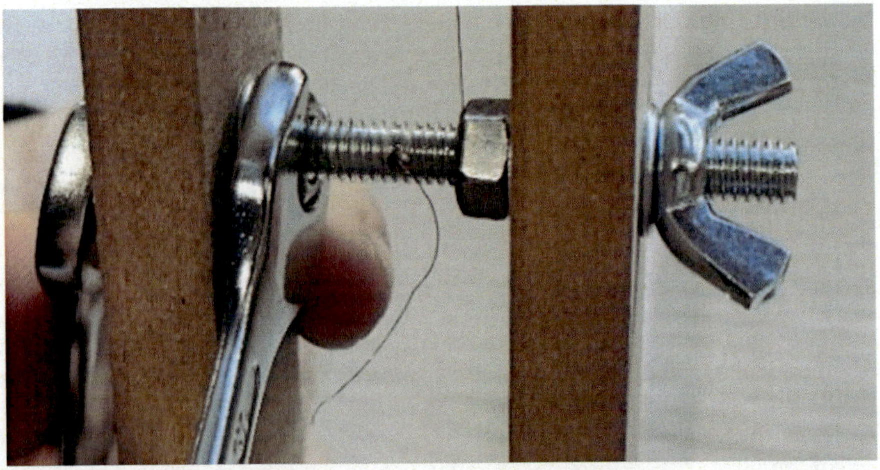

Fig. 12.10 Thread tension adjustment

12.4 Job Description

Each workstation includes the following elements:

- Adjustable power supply with 2 connectors (at the output of the power supply the voltage must be 24 V).
- Metal arch with four wooden arms
- 2 airfoil intrados contours
- 2 extrados contours of the aerodynamic profile
- 2 metal brackets

Fig. 12.11 Full cutting arch position

- 2 fixed wrenches of the metric of the screws
- 1 spool of nichrome yarn

To start up the cutting wire it is necessary:

1. Connect the power supply to each side of the cutting wire: connection end to the end where the spring is located and the end where the wing nut is located.
2. On the voltage source: The voltage to be set will vary according to the wire diameter and length, as these parameters will vary its resistance. For the selected wire (0.5 mm) and a wire length of 1200 mm, the tension will be set between 12 and 18 V. The wire should be checked to ensure that it is hot enough to cut the extruded polystyrene sheets properly.

12.5 Procedure for the Realization of the Practice

The procedure to carry out the practice will be as follows:

1. Polystyrene (XPS) sheets with sufficient thickness to cover the profile guides shall be used (if the available sheets will not reach the minimum thickness, two would be glued). Then, with the power supply switched off, the poles shall be placed at the ends of the arch: on the spring washer and on the dowel screw.
2. The cutting area shall be marked on the plates. This area will be defined by the length of the cutting guides and the wing span, forming a rectangular area. The cutting area is marked on the plates. This area will be defined by the length of the cutting guides and the wing span, forming a rectangular area. At the end of the cut, again, the power supply is switched off.

Fig. 12.12 Preparation for cutting the lower part of the guideway: intrados

The intrados cutting guides will be placed on the sides of the resulting block, so that the leading edge of the guide is perfectly aligned with the edge of the block. It is very important that the guides are well glued to the block with a contact adhesive spray or double-sided tape, so that they do not come off during cutting.

The intrados guide (see Fig. 12.12) is cut with hot wire, with one student on each side of the arch holding the airfoil and guiding the wooden arms along the guide. It is very important that the cutting speed was low and that the arch follows the guide perfectly at all times, as any deviation will cause defects in the core. For this purpose, it is recommended to lay the arch on the table and simply guide the arms on each side. When the cut is finished, it is mandatory to turn off the power supply to avoid burns.

Once the intrados guides (¿guides of the lower profile?) have been cut (see Fig. 12.13), remove the guide and place the top guides (¿guides of the upper profile?) in the same position (see Fig. 12.14), repeating steps 4 and 5 for the top guides. Again, it is very important not to deviate from the profile contour and to turn off the power supply after each cut.

Finally, the foam guides are removed and the core is extracted from the interior of the foams for future use. The remaining volume will be the female core (the geometry defined by intrados and extrados, as shown in Fig. 12.15) and can be useful for the manufacture of hollow profiles by laminating these contours.

Fig. 12.13 Result of the intrados cut

Fig. 12.14 Upper part of the guide: top surface

Fig. 12.15 Definition of the cut profile (result and superimposition of the guides)

12.6 Conclusions

With this practical session, students work on technical aspects covered in different subjects during their undergraduate studies and also learn to work in a team.

From a technical point of view, the practical was a success as the students were able to apply their technical knowledge of aerodynamics to manufacture airfoils and a hot-wire cutting arch.

From a practical point of view, the students were able to handle the tools to manufacture the necessary components to obtain a polystyrene spoiler.

There is no doubt that developing a practice of this type, where students must generate their own tools and use them to obtain a product, greatly enhances their motivation and learning.

References

1. Chenglie, L.: The Story of Confucius. Editorial Lectorum, S.A. (2006)
2. Enzo, F.: The Enzo Ferrari Story. The Macmillan Company, New York (1964)
3. Meederira, P.B.: Aerodynamic development of a IUPUI Formula SAE specification car with Computational Fluid Dynamics (CFD) analysis. Ponnappa Bheemaiah Meederira, Indiana, University Purdue, University. University Indianapolis (2014)
4. NACA Profiles. www.despegamos.es. Accessed 25 Mar 2022

Chapter 13
Plans for a Course on the History of Mechanisms and Machine Science

Marco Ceccarelli and Marco Cocconcelli

Abstract Many universities offer optional courses within the curricula of studies that students can choose based on their interests. Usually, they are short courses of 3 ECTS (credits) but in some cases they can be proposed up to 6 ECTS. Moreover, they are often common to several degree courses, so the topics covered should be more general and transversal with respect to the specific engineering curriculum. In this paper, the background significance of the History of Mechanisms and Machine Science (MMS) is discussed by re-proposing a short course in technical formation curricula for engineers, preferably at Bachelor levels. After reviewing some previous preliminary experiences, the course proposal is outlined as based on the expectations in learning outcomes and with a general structure referring to basic literature. The target is to provide historical education backgrounds within the formation curriculum of a modern engineer.

Keywords History of mechanisms and machine science · Education · Course Synopsis

13.1 Introduction

The curriculum of any engineering degree is fulfilled with technical courses giving solid foundations about scientific and technical matters like calculus, physics, hydraulics, dynamics. In fact, the aim of any STEM (Science, Technology, Engineering e Mathematics) course is to prepare the student to be guided by the so-called

M. Ceccarelli (✉)
Laboratory of Robot Mechatronics, Department of Industrial Engineering, University of Rome Tor Vergata, Via del Politecnico 1, 00133 Roma, Italy
e-mail: marco.ceccarelli@uniroma2.it

M. Cocconcelli
Department of Sciences and Methods for Engineering, University of Modena and Reggio Emilia, Via G. Amendola 2-Pad., 42123 Morselli, Reggio Emilia, Italy
e-mail: marco.cocconcelli@unimore.it

J. C. García Prada et al. (eds.), *Trends in Educational Activity in the Field of Mechanism and Machine Theory (2018–2022)*, Mechanisms and Machine Science 128,
https://doi.org/10.1007/978-3-031-25730-8_13

"scientific method" that is the way in which science proceeds to reach a reliable and verifiable knowledge of reality. Although most of the students have a clear idea of it, very few of them know it was formalized by Galileo Galilei in 1593 [1, 2]. In academic courses, the scientific results are not contextualized in their historical period, due to a lack of time or the idea that is a negligible information compared to its technical implications. Nevertheless, the history of the evolution of the machine theory and its consequence in technical applications and everyday life tools are fundamental for a proper understand of the contribution of distinguished researchers and the importance of the technical contents in an engineering course. In general, the history of the society is studied by referring to facts and characters in the political and even military fields, also considering them as a condition and cause of the relative changes and evolutions also in the scientific and technological environment. But alternatively, it can be thought that conditions and evolutions in science and technology have determined those conditions, and political and even military changes have determined these historical events. This interpretation is certainly plausible in the distant past and with reference to machine mechanics, and nowadays even more evident.

The evolution of technology has always followed a non-linear growth and in recent years we have seen great changes in a few years. For example, the advent of machine learning fastened the developed of autonomous driving that sounded like science fiction until few years earlier. Similarly, these accelerations in the technical developments also occurred in the past, for example during the Renaissance or after the Second World War. A historical framework of the technical evolution allows to better understand the importance of the figure of the engineer himself/herself [3], who is the architect of this process, of the relations with the political and social context of the time and, consequently, to have a greater awareness of his/her function in the society.

On a practical way, the collection of structured information about inventors and inventions has also an educational aspect. To name a few: the difficulty in finding sources that may be scarce or difficult to obtain (think for example of old or out-of-print books, or short-run publications no longer available), the problem of verifying the sources, of contextualizing in the historical period and finally of interpreting them in a structured report. Once assimilated, these skills that we could consider as soft skills, could be applied in different areas during the student's career.

So far, classes on History of Technology within an engineering course are not present in Italy. Few exceptions were the courses proposed by at both Bachelor and Master level but also as a monographic course for Ph.D. students by teachers interested on History of Engineering by using their fundings and with sporadic and optative offer, mainly in the years 2005–2015, like for example the author experience [4]. Other specialized courses are proposed in the degree courses in History, but they designed for schools of humanities rather than engineering students. However, it is important that the lecturer has a technical background rather than just a historiographical one, to highlight the technological aspects of inventions and make the course more meaningful for engineering students.

Starting from these premises, in this paper, the authors propose the structure of an academic course on the History of Mechanisms and Machine Science (HMMS).

The target is a class of a bachelor engineering students, who will learn in a 3 ECTS course the milestones and distinguished figures who contributed to the technological level of everyday life. The paper is structured as follows: the motivations for a such a course and the feedback from past experiences of the authors are given in Sect. 13.2, while Sect. 13.3 details the outlines of the course in the frame of Bologna Process outcomes.

13.2 Motivations and Previous Experiences

The training activity that is proposed with a course on the history of engineering for engineering students aims to give students and therefore young engineers a historical awareness of identity of their own professional cultural environment with historical references that recognize evolutions and useful characteristics to have a conscience, a pride, and a vision of the importance and role that an engineer can have for the technological, social and cultural development of society, even if with her/his activity not expressively aimed at such ambitious goals.

Figure 13.1 summarizes in a synthetic way a wide breadth and variety of motivations and areas of interest that can motivate and justify the three main aspects of an introductory course on the history of engineering and machines with content in the three main aspects of culture historical with a general outline, a historical awareness of the machines and inventors/designers of the past, and a capacity for historical-technical analysis of the achievements of the past in machines and in technical-scientific-professional profiles. The need or rather the convenience of having a broad basic culture, including historical ones, is certainly a precious baggage for a complete vision of how much impact an engineering product has had and can have in the future. In addition to the tu-court cultural aspect, a specific awareness of the past in the technical-scientific field of own interest can be a valid foundation for a clear vision of future developments, also for her/his own activity. These two considerations can be the motivation for having a general section of the course on the history of engineering and also on the mechanical one. The knowledge of the past of machines is certainly based on the knowledge of the machines of the past and of the inventors and designers, who have determined their technological success but also an impact on society, giving the student the opportunity to have historical examples but a reference of how much the development of a machine or an invention requires and how much it can be useful in the technological and social fields. Not wanting to propose a course merely in the history of engineering, which is typical of the humanity fields of the history of the science and technology, the course can be aimed at the interest and application of an engineering student in the formation of practical skills in the search for original documents, also of a literary type, with the relative analysis and interpretation and in the ability to use specific historical-technical analysis tools of the characteristics of the machines of the past and of the historical evolution of the activities of the inventors and designers of the past.

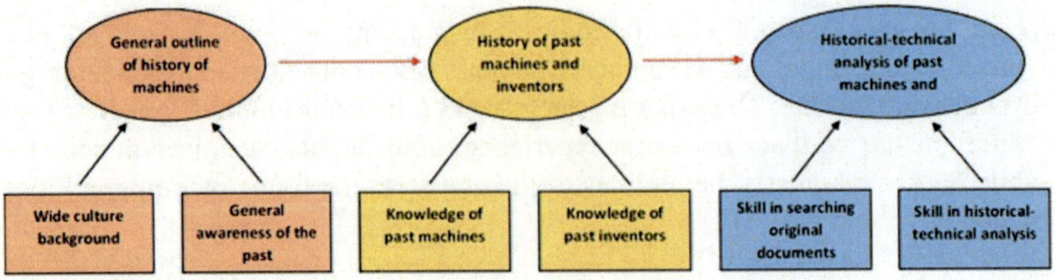

Fig. 13.1 Motivations versus contents for a basic course on History of MMS

This thematic and motivational plan of the course can be summarized in a schedule of lessons, also of a synthetic and even point-to-point type with study indications, as reported in Table 13.1 that is extracted from a previous experience of one of the authors in a course held in the year 2008–2013, [4]. It must be noted that the history of technique and engineering, as outlined in part in Fig. 13.1, is well addressed and documented in wide publications such as in encyclopedic works as for example [5, 6], in which the machines are presented as a part of technological development although also with specific references, in monographs such as those in [7, 8] in which the role and importance of machines in the historical development of technology in general as well as of society is specifically examined, and in specific paper publication like [9] in which the teaching significance and cultural heritage of past mechanism models is discussed as for research findings. These interests are also addressed in congress events where the primary role of machines and its inventors is recognized as a fundamental and priority issue in research and interpretation activities as reported in the proceedings of the congresses of the HMM symposium, [10], that is organized periodically as part of the activity of the IFToMM Permanent Committee for History of MMS. A particular attention on the technical-scientific profiles is dedicated to the works that are recently collected in the Distinguished Figures series, [11], where the personality of an inventor is analyzed not only from a biographical and historiographic points of view but also according to the technical-scientific activities that have characterized her/his profile becoming a reference in technology and in the history of machines.

The here-in proposed course also requires a concrete application of the hints of the history of the machines offered to the students also as a background analysis of new designs and new developments. Figure 13.2, [12], shows an example of this activity carried out within the course with the previous experience of one of the authors, [4], with characteristics such that the historical-technical analysis carried out by the student was worthy of publication as well as in [13]. This example shows how actually knowing the historical-cultural background with the technical-scientific characteristics can give the possibility to understand a product with the possibility of further innovative developments with an appropriate awareness of the proposed innovation.

Table 13.1 Example of previous experience: a plan of lectures

Evolution of the term and status of engineers; historical period and definitions
Tutorial: retrieval of historical document and choice of essay theme
Personality and Mechanism Design in Antiquity (Greco-Roman)
History of mechanics applied to machines—examples of personality
Evolution of the classification of mechanisms
Tutorial: study or design interpretation of the past
Medieval technical culture and knowledge transfer; Personality and design of mechanisms in the Middle Ages
Tutorial: dissertation elaboration
Culture and machines of the renaissance; analysis and applications of mechanisms in the Renaissance
Tutorial: dissertation elaboration
History of the theory of helical motion
Machines and mechanisms in the Industrial Revolution: development of types of mechanisms
Tutorial: thesis elaboration and presentation

13.3 Outlines for the Course

The proposed course is made up of three complementary activities referring to Fig. 13.1. The first is the core part of the course that is the general outlines of the history of mechanisms and machine engineering starting from Prehistory till the twenty-first century. It is advisable to make use of many images during the lesson, with the aim to encourage the involvement of students but also to show the evolution of technical documentation over the centuries. An example is the interesting book [7] with an illustrated history of HMM that can be suggested as reference textbook of the course. The second activity is complementary to the main one, analyzing how a specific machine or mechanism works with analytical tools that students already know from other courses, or presenting the historical and technical analysis of the scientific work of relevant inventors. An example of literature of reference is the series of books [11] on the distinguished figures in MMS which collects contributions on 53 relevant figures starting from Antiquity to the present day divided over 4 volumes up today. The last activity, interactive with student, is a practice on how to carry out, i.e. research and report, a historical-technical analysis. Starting from examples given in the second activity, the students can choose a topic (a scientific discovery or invention) and develop an historical-technical analysis by their own, regularly supervised by the teacher.

The historical developments of mechanisms and machines as related to mechanism design can be divided into periods with specific technical developments that,

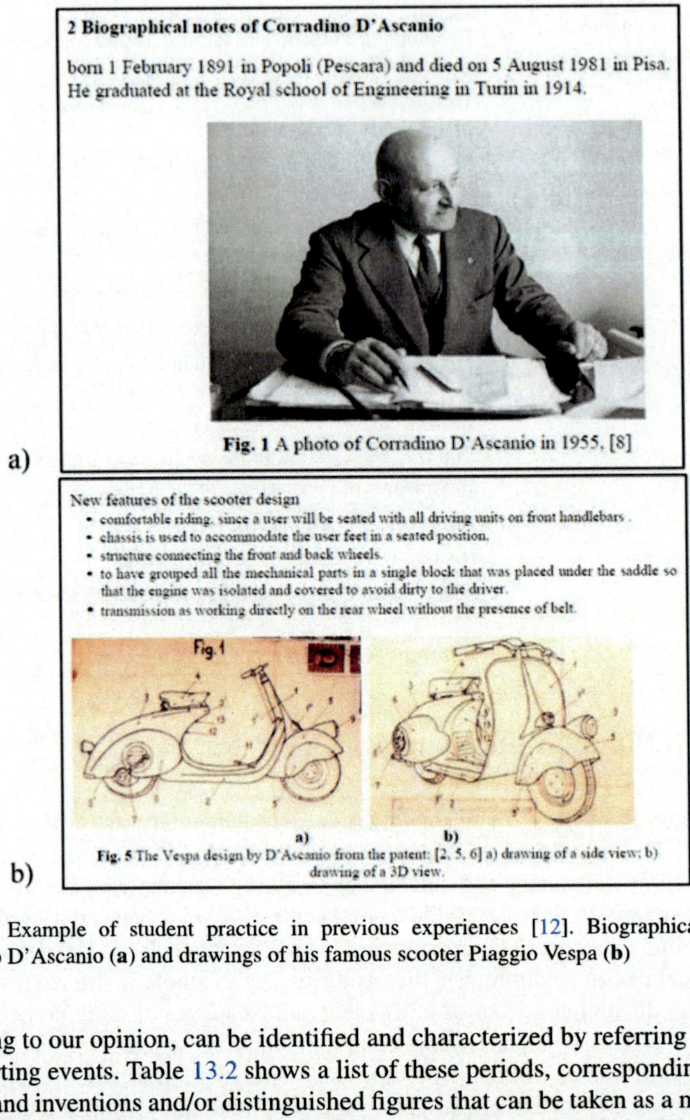

Fig. 13.2 Example of student practice in previous experiences [12]. Biographical notes of Corradino D'Ascanio (**a**) and drawings of his famous scooter Piaggio Vespa (**b**)

according to our opinion, can be identified and characterized by referring to significant starting events. Table 13.2 shows a list of these periods, corresponding starting events, and inventions and/or distinguished figures that can be taken as a main reference of the course structure (for the first two activities). A detailed analysis of the synopsis of the course is out of the scope of this paper due to the limited space available.

Figure 13.3 shows the flowchart of the practice activity, i.e. the homework that the students have to develop during the course and that will be presented at the final exam for evaluation. The activity is divided into four main parts: preliminary steps, documentation, interpretation, and conclusive presentation. In the preliminary step the class is divided into teams of 2–3 students and each team must choose a topic for their study (e.g. a machine, an invention, a theory, a technical production, an inventor) and to propose a tentative title. The second part is devoted to the search for

Table 13.2 Suggest historical periods for HMMS and starting events

Period	Year	Starting event/Invention/Inventor
Prehistory		Utensils
Antiquity	5-th century BC	Mechanos in Theater plays
Middle Ages	275	Sack of the School of Alexandria
Early design of machines	1420	Zibaldone by Filippo Brunelleschi
Early discipline of mechanisms	1577	Mechanicorum Liber by Guidobaldo Del Monte
Early Kinematics of mechanisms	1706	Traitè des Roulettes by Philippe De La Hire
Beginning of TMM	1794	Foundation of Ecole Polytechnique
Golden Age of TMM	1841	Principles of Mechanism by Robert Willis
World War Period	1917	Getriebelehre by Martin Grubler
Modern TMM	1959	Synthesis of Mechanisms by means of a Programmable Digital Computer by Ferdi-nand Freudenstein and Gabor N. Sandor
MMS age	2000	Definition of Mechanism Science

documentation (original documents, references, using physical or online supports) and their analysis in terms of sources and bibliography. The third part is probably the most important one: the student's original elaboration of the material found. The activity may consist in a brief analysis of the historical context, the evaluation of the impact of the invention in that historical moment, a technical analysis of the novelty of the invention and its valorization recognized at that time, a technical analysis of the invention from a modern point of view (modern techniques approach, numerical performance analysis, etc.). Finally, in the fourth part the work is summarized in a short report (max 30 pages) with illustrations, graphic representations, numerical tables, references, and in an oral presentation on with the participation of all the members of the team.

Table 13.3 lists the expected learning outcomes of the proposed course according to the current rules for teaching planning [14]. It is intended as a reference, in the sense that each lecturer can customize it depending on the peculiarities of the specific engineering course (e.g. in the context of an electrical engineering on relevant figures in the development of electric motors, while in mining engineering more attention will be devoted to hydraulic contributions).

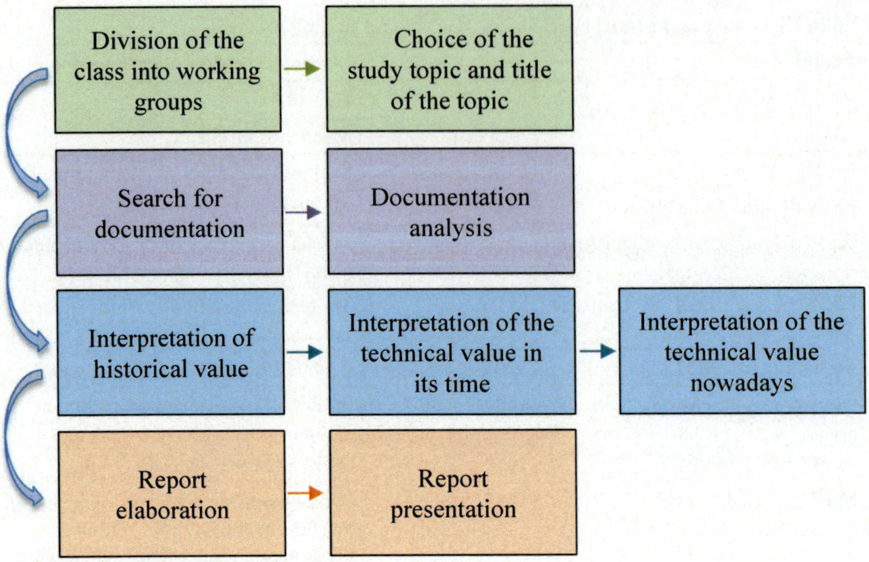

Fig. 13.3 Flowchart of the practice activity

Table 13.3 A workplan and offer of the proposed course

Course name	Fundamentals of history of mechanisms and machine science
Level	1 (bachelor's degree)
ECTS	3–6
Educational goals	The course focuses on an outline of the History of Mechanisms and Machine Science with examples. The description of the main technical and technological advances from prehistoric times up to the present day, focusing on main distinguished figures who significantly contributed to the development of machine science
Prerequisites	There are no formal restrictions on access to the course, however foundations on Mechanics of Machines are certainly cultural technical prerequisites
Contents	See Table 13.2
Teaching methods	Teaching is provided in English through face-to-face lectures. The lessons include a theoretical part and a part of homeworks. The theoretical part is aimed at the understanding and deepening knowledge of the topics covered by the program. The homeworks aim to form the student's skills in searching and understanding technical contributions in the History of Machine Science
Verification of learning	The final exam is divided into two parts. Two weeks before the exam date, the students should send a report of their individual homeworks. The students are evaluated both on the base of the report provided, and an oral interview on the whole program

(continued)

Table 13.3 (continued)

Course name	Fundamentals of history of mechanisms and machine science
Textbooks	Main textbooks on History of Mechanical Engineering
Expected learning outcomes	(1) Knowledge and understanding – Knowing and understanding the main technical and technological advances in the history of mechanisms and machine science – Knowing the main distinguished figures that contributed to the development in the field of engineering (2) Applying knowledge and understanding – Knowing how to date the historical period for a given mechanisms or a machine – Knowing how to find and set historical documents and details regarding a mechanism or a machine (3) Making judgements – Ability to date a mechanism or a machine – Ability to retrieve historical details for a mechanism or a machine of interest – Ability to evaluate the past and present value of past achievements (4) Communication skills – Knowing how to integrate the knowledge of a device with historical information regarding the history and evolution of the technology used (5) Learning skills – Ability to continuously update the techniques and methodologies used in a professional context

13.4 Conclusions

The paper introduces motivations for a short course on History of Machines and Mechanisms within a formation of a modern engineer, preferably at Bachelor level with the aim to provide a basic culture and awareness of the past. A content of the course is outlined in three main parts, namely general History outline, examples of part machines and inventors, and a procedure for a practice on a historical-technical analysis.

References

1. Galilei, G., Mecaniche, L.: Opere di Galileo Galilei (F. Brunetti (ed.)). UTET, Turin (1964)
2. Ceccarelli, M.: Early TMM in Le Mecaniche by Galileo Galilei in 1593. Mech Mach Theory **41**(12), 1401–1406 (2006)
3. Ceccarelli, M., Bragastini, R.: Historical accounts on the figure of engineers and academic mission for their formation. In: New Trends in Educational Activity in the Field of Mechanism and Machine Science, pp.3–10. Springer, Dordrecht. https://doi.org/10.1007/978-3-319-01836-2_1, ISBN 978-3-319-01835-5

4. Ceccarelli, M.: Lecture notes of short course on history of mechanical engineering. University of Cassino, Cassino (2012)
5. Capocaccia, A. (ed.): UTET, Turin (1973)
6. Singer, C., et al. (eds.): Boringhieri, Turin (2012)
7. Paz, E.B., Ceccarelli, M., Otero, J.E., Sanz, J.L.M.: A Brief Illustrated History of Machines and Mechanisms. Springer, Dordrecht (2010)
8. Koetsier, T.: The Ascent of GIM, the Global Intelligent Machine-A History of Production and Information Machines, History of Mechanism and Machine Science. Springer, Cham
9. Ceccarelli, M., Cocconcelli, M.: Italian historical developments of teaching and museum valorization of mechanism models. Machines **10**(8), 628 (2022). https://doi.org/10.3390/machines10080628
10. Ceccarelli, M., et al. (eds.): Explorations in the History and Heritage of Machines and Mechanisms- International Symposium on History of Machines and Mechanisms (HMM). History of Mechanism and Machine Science, 7 vols. Springer, Cham
11. Ceccarelli, M., et al. (eds.): Distinguished Figures in Mechanism and Machine Science - Their Contributions and Legacies. History of Mechanism and Machine Science, 4 vols. Springer, Cham
12. Teoli, G.: Practice report: Corradino D'Ascanio and His Design of Vespa Scooter, short Course on History of Mechanical Engineering. University of Cassino, Cassino (2013)
13. Ceccarelli, M., Teoli, G.: Corradino D'Ascanio and His Design of Vespa Scooter, Multibody Mechatronic Systems, Mechanisms and Machine Science, vol. 25, pp. 399–410. Springer, Dordrecht. https://doi.org/10.1007/978-3-319-09858-6_38
14. Gudeva, L.K., Dimova, V., Daskalovska, N., Trajkova, F.: Designing descriptors of learning outcomes for Higher Education qualification. Procedia – Soc. Behav. Sci. **46**, 1306–1311 (2012)

Chapter 14
A European Researchers' Night Project on Mechanical Vibrations for High School Students

Marco Cocconcelli, Cosimo Fonte, Pasquale Grosso, Giovanni Mottola, Matteo Strozzi, and Riccardo Rubini

Abstract The present works were conceived to be exhibited during the 2022 European Researchers' Night (ERN 2022), at the University of Modena and Reggio Emilia. The idea is to illustrate the key concepts of mechanical vibration through the use of 3D models and virtual simulation analysis. The paper is directed to high school students planning to enroll in a mechanical engineering bachelor's degree, in order to approach or consolidate some fundamental concepts of mechanical vibration. Topics not easy to explain, such as the natural frequencies of a body, could be presented more effectively using physical models. Mathematical formalism will be kept to a minimum, as it is beyond the scope of this paper.

Keywords Mechanical vibration · Natural frequencies · Modal analysis

14.1 Introduction

Any motion that repeats itself after an interval of time is called vibration or oscillation [1]. Two main types of vibrations are generally defined: *forced* and *free*. A *forced vibration* happens when the system is subjected to an external (usually periodic) force. On the other hand, if a system, after an initial disturbance, vibrates without external variable forces, one has a *free vibration*: this introduces the concept of *natural frequency*. The study of natural frequencies is extremely important because when bodies are excited at their natural frequency, a resonance occurs, which leads to high deformations of the structure [2]. Each natural frequency of a body is associated with a shape, a way in which the body vibrates, called *natural mode*. In the next sessions the natural modes and frequencies for three different structures will be identified. The first (Sect. 14.2) is a simple case of cantilever beams with attached point masses, whose vibrations are manually excited. The second structure (Sect. 14.3) is used to

M. Cocconcelli (✉) · C. Fonte · P. Grosso · G. Mottola · M. Strozzi · R. Rubini
Department of Sciences and Methods for Engineering, University of Modena and Reggio Emilia, 42122 Reggio Emilia, Italy
e-mail: marco.cocconcelli@unimore.it

J. C. García Prada et al. (eds.), *Trends in Educational Activity in the Field of Mechanism and Machine Theory (2018–2022)*, Mechanisms and Machine Science 128,
https://doi.org/10.1007/978-3-031-25730-8_14

represent the modes of a three-storey building, excited with a harmonic force generated by a mechanical shaker, to show how these vary using a mass damper. In the last example (Sect. 14.4), a plate clamped at its center is adopted as object on which perform an experimental modal analysis, identifying some natural frequencies and showing mode shapes through software interface. All the examples were first explained in the simplest way, then the experiments were performed in order to visualize and convey the concepts in a more engaging and stimulating way.

14.2 Vertical Beams with Point Masses

The first model developed is a small cart with three vertical beams, each carrying a concentrated mass along its length (Fig. 14.1). This concept was inspired by similar models [3], in which a mechanical system is designed to display several resonances at low frequencies. This way, the vibrations can be manually excited, and it is easy

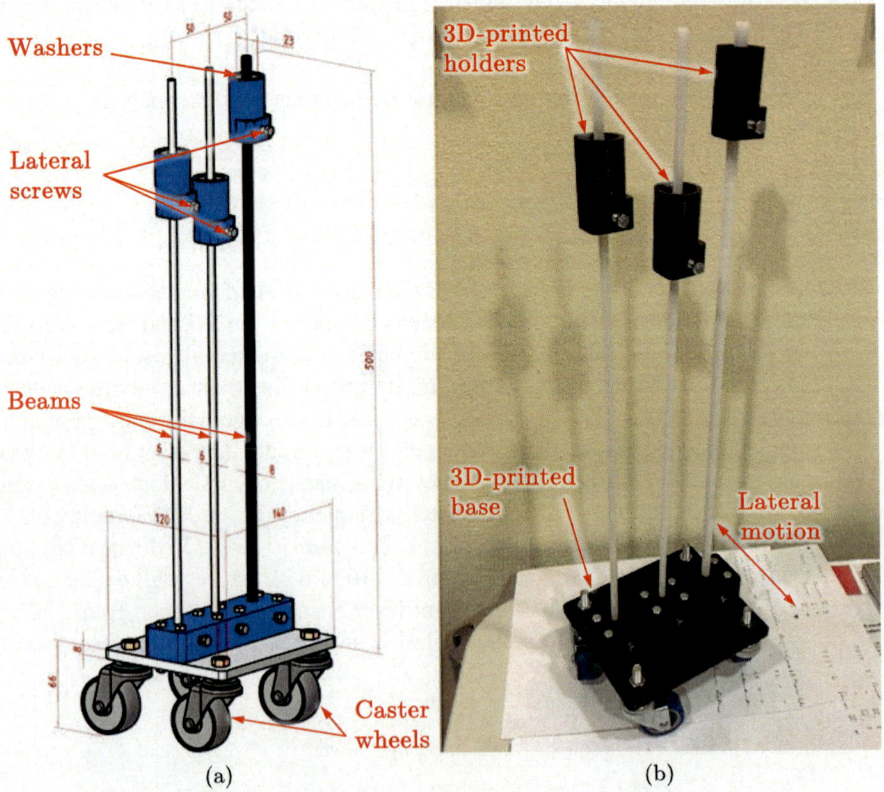

Fig. 14.1 **a** The CAD concept we developed. **b** The final built model of the cart

to distinguish the behavior at different frequencies. Our design has the following features, which we used as guidelines in this project.

1. *Configurability*: the design can be easily modified during a presentation to change the dynamic behavior of the structure. For simplicity, we consider a simplified model of a slender cantilever beam (with distributed mass M_d) vibrating with small displacements and carrying a point mass M_c at a given position a along its length L. The first natural frequency can be found analytically, but it requires solving a complex trascendental equation [4, 5]. This equation depends on L, a and M_c, on the cross-section properties of the beam (the area S and second moment of area I) and on the intrinsic properties of its material (namely, its stiffness modulus E and density ρ). Thus, the design allows to vary all these parameters independently:

   – the lengths a and L can be changed by unlocking lateral screws that fix M_c along the beam and the beam on the cart, respectively;
   – the mass M_c can be changed by adding or removing washers on the 3D-printed holders that are placed on each beam;
   – the beam properties can be changed by using beams of different materials (in our tests, we used plastic bars composed of POM, PVC and Nylon).

   The cart has three slots for placing the beams, two with a diameter of 6 mm and one with a diameter of 8 mm, so the effect of changing the cross-section can also be evaluated (by comparing the behavior of the three beams).
2. *Transportability*: the model can be disassembled and reassembled easily, for ease of use in classroom activities and at public events such as ERN.
3. *Effectiveness*: the design use as cheap elements found in hardware stores, with some 3D printed parts realized with a standard FDM printer.

The cart is manually excited, by moving it back and forth on a flat surface; the caster wheels on its base allow for quick and frictionless motion. It can then be seen how one of the beams (the one whose natural frequency is closer to the excitation frequency) vibrates with higher amplitudes: the demonstration is particularly effective if the movable masses are placed such that the beams have natural frequencies that are well apart from each other. A metronome can be used for keeping the rhythm; the experiment can also be designed as a teaching game, where the students have to find the right frequencies by trial and error.

We also developed an Excel worksheet to compute the first natural frequency of the beams: besides being useful in the design phase, this worksheet can be an interactive teaching tool, to compare the expected frequencies with the real ones.

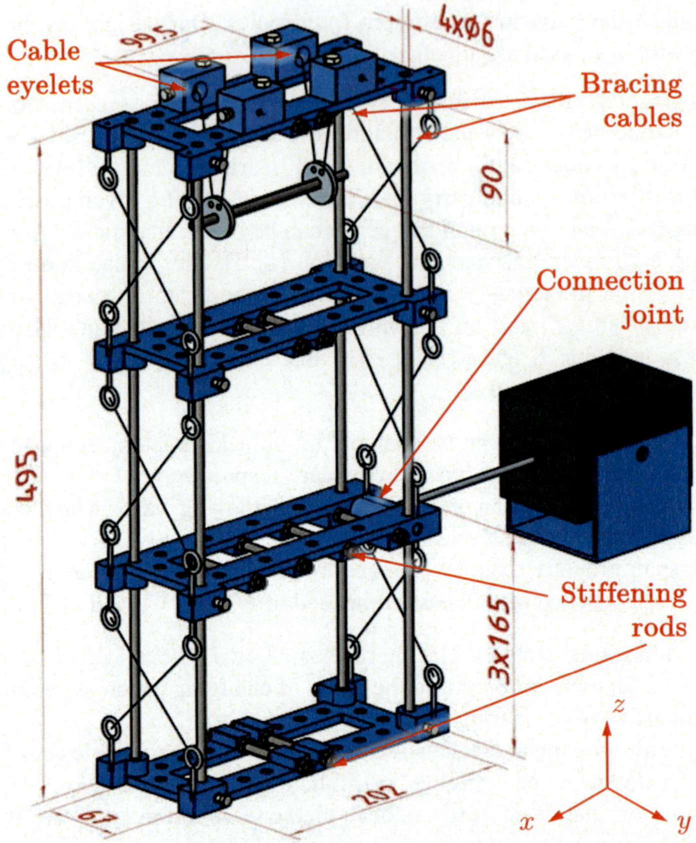

Fig. 14.2 CAD model of the building; dimensions in mm. The shaker is also shown

14.3 Vibrating Building with Tuned Mass Damper

The second structure is a model of a three-storey building vibrating under a lateral load; the CAD concept is shown in Fig. 14.2. While other similar demonstration tools exist [6, 7], this model was designed as a configurable and inexpensive tool, like the one presented in Sect. 14.2: therefore, the same design guidelines were followed in this case, too. Thus, the floors can be moved along the length of the four columns (composed of 6 mm POM beams); their masses can also be tuned by adding or removing nuts that can be clamped on the available holes.

This design uses lateral wires for bracing the structure: this way, the shear motion (along the x axis of the coordinate frame, which is aligned with the axis of a shaker) corresponds to a relatively low stiffness, while other motions (such as bending or torsion of the whole structure, axial displacements along z or shear motion along y) have much higher stiffness. Since the shaker operates at low frequencies

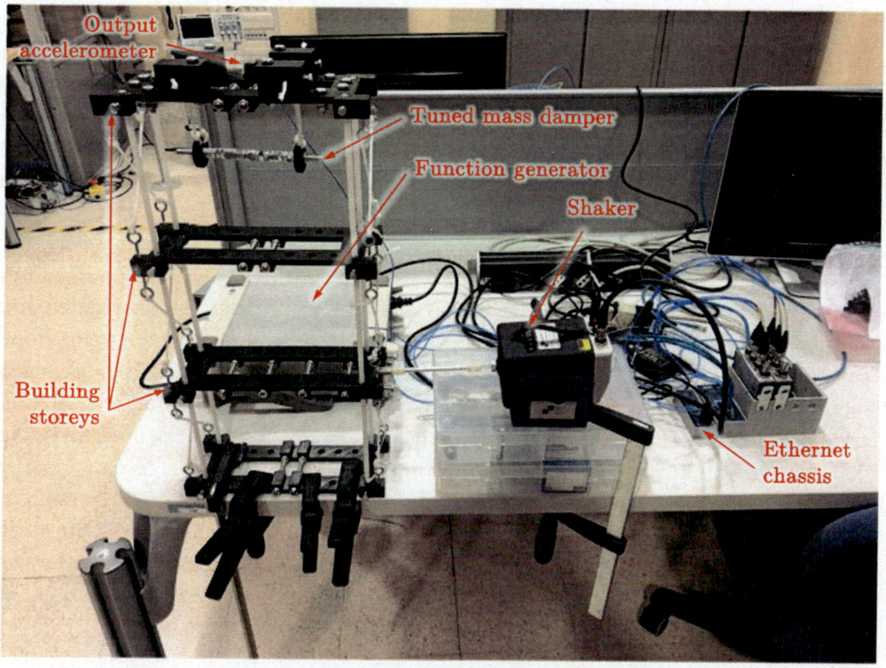

Fig. 14.3 Photo of the vibrating building clamped on the test table for modal analysis

in this case, only the natural modes corresponding to shear along the x axis are of interest. We also use stiffening rods, to reinforce the floors (composed of 3D printed plastic and thus relatively flexible); this way, they can be considered as rigid masses. In any case, it was also verified that the model can be used without these wires and rods, to tune its stiffness as desired.

The model is connected to the shaker through a rotary joint on the first floor: this was required to avoid misalignment of the shaker connection. In this way, one can also lower the shaker with respect to the model and verify the effect of having both horizontal (shear) and axial excitation forces. For simplicity, the shaker axis was kept horizontal in the tests, by raising the shaker base.

The final built model is in Fig. 14.3. The frequencies of this building may be found through analytical methods [8, 9], by considering the structural compliance to be concentrated in the columns. This, however, proved to be complex due to the uncertainties in the structural parameters, such as the properties of the material used for the columns. Instead, an experimental modal analysis was performed, by exciting the structure with an impulse (which is produced by a function generator) and measuring the vibrational response. Two piezoelectric accelerometers (with sensitivities of 10 mV/g) were placed on the model: one close to the shaker connection point, measuring the input excitation, and the other one on the top floor, which is expected to have the largest oscillations. Three-axial accelerometers were used, in which the channels of interest are the axes of measurement aligned with the x axis; the accelerometers were connected to a NI cDAQ Ethernet chassis sampling at 51,200 Hz. By computing the FRF between the input and the output channels, the natural frequencies of the structure were found at 1, 6.5 and 24 Hz. By exciting the system at these frequencies with sinusoidal forces (again by controlling the function generator), the vibration modes can be clearly shown, since the displacement of each floor visibly changes.

This model was also designed to be used with a tuned mass damper, which can be rapidly connected and disconnected to show the different behavior in each case. This is composed of a pendulum suspended on the third floor: while a simple point mass attached to a rope could be applied [10], it would be difficult to constrain its motion along the x axis. Therefore, in our design, we used a threaded bar suspended at its ends by four cables (which pass through cable eyelets on supports fixed to the floor): since the cables are of the same length, the bar always remains horizontal and no inertia torque is introduced. Thus, its behavior is dynamically equivalent to that of a simple pendulum: if an excitation is applied at the frequency of the pendulum (which depends only on the cables' lengths), the energy introduced is in large part absorbed by the pendulum, which swings with large amplitudes, while the vibration on the rest of the structure is reduced. This was verified experimentally, by measuring the output vibration both with and without the damper: the RMS amplitude of the acceleration is reduced by $>10\%$. This effect can be varied by changing the mass of the pendulum (on which nuts can be screwed), without changing its natural frequency.

14.4 Plate with Center Point Clamped

14.4.1 Setup

This section presents the setup used to carry out the experimental modal analysis of a square plate with the central point clamped. As shown in Fig. 14.4, the plate was constrained in its central point by interlocking to a massive vertical support bar. Being interested in out of plane modes, a uniaxial accelerometer with sensitivity of 10 mV/g was fixed on the lower face of the plate with vertical acquisition direction, while the impulses were supplied by an instrumented hammer with a sensitivity of 2.25 mV/N with a metal tip. Both devices were connected to the acquisition system case through BNC connectors cables. Siemens equipment, LMS SCADAS Mobile (hardware) and LMS Test.LAB (software) were used for acquisition and analysis respectively.

14.4.2 Measurements

The plate ($200 \times 200 \times 3$ mm) was discretized giving a regular grid of 121 points to be excited. The measurements were carried out by exciting the structure with 5 impacts per point, averaged in *coherence* and *frequency response function*.

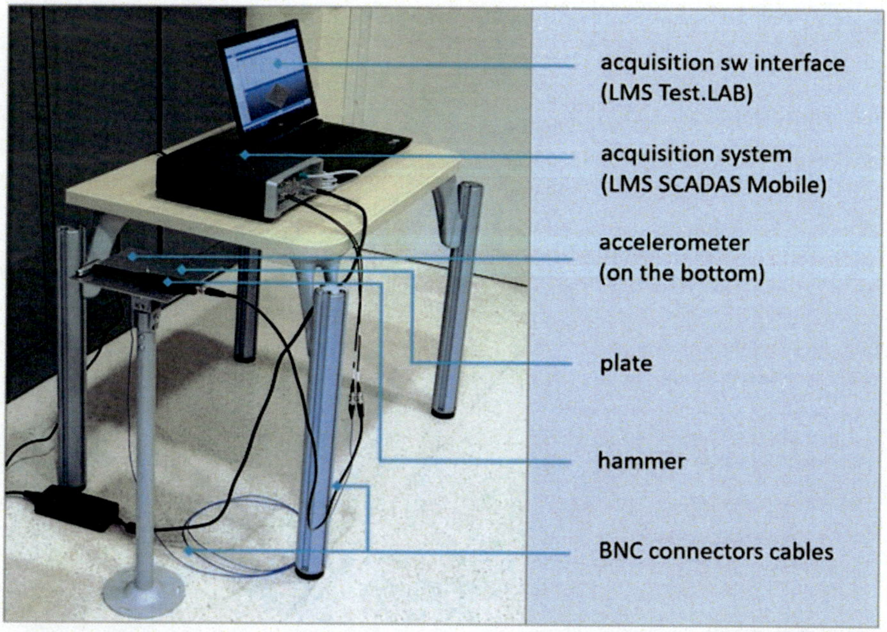

Fig. 14.4 Setup

14.4.3 Results

Starting from the complete set of acquisitions, in the post-processing phase a global frequency response function was calculated in order to analyze all the data simultaneously, and thanks to the Least Square Complex Exponential method, estimate values of natural frequencies and related damping percentage has been obtained (see Table 14.1). The range covered is up to 1000 Hz, due to limits related to impact energy supplied manually, and a good measurement accuracy is also required. The corresponding modal shapes has been finally obtained, Fig. 14.5 shows 2 of the 8 identified modes: for each of them, the maximum amplitude deformations in opposite phases are shown, nodal lines are marked in red.

14.5 Conclusions

The different projects define a course in mechanical vibration analysis, starting from systems with a single degree of freedom (Sect. 14.2) and moving to models with many degrees of freedom that better approximate a continuous body (e.g. the third project). Positive and full of admiration feedback have been received

Table 14.1 Natural frequencies detected [0–1000 Hz]

Mode	Frequency [Hz]	Damping [%]
1	63.259	1.42
2	190.424	2.05
3	230.459	0.65
4	339.916	0.49
5	542.651	4.50
6	601.901	0.75
7	750.183	2.24
8	997.938	1.10

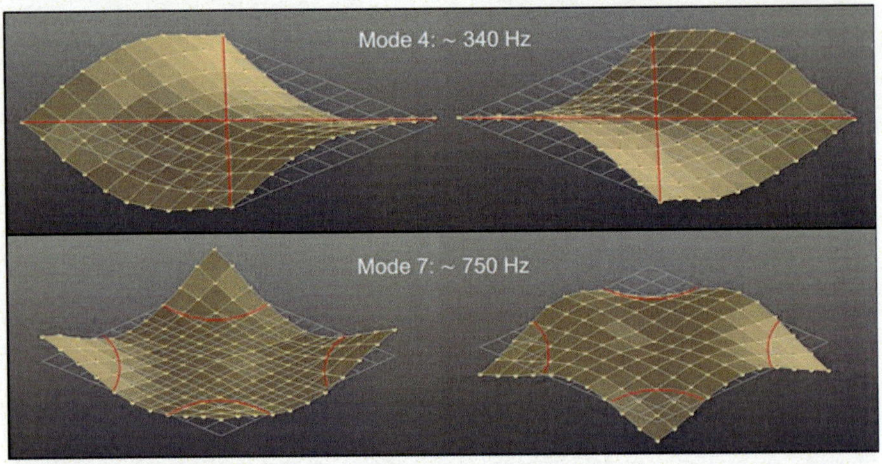

Fig. 14.5 Modal shapes

from high school students, who, driven by a strong interest in mechanics, have expressed their desire to enroll in a degree course. In particular, they appreciated how the use of physical models is fully effective in understanding complex phenomena.

References

1. Rao, S.S.: Fundamentals of vibration. In: Mechanical Vibration, sixth edn. Pearson (2017)
2. Green, D., Unruh, W.G.: The failure of the Tacoma bridge: a physical model. Am. J. Phys. **74**(8), 706–716 (2006). https://doi.org/10.1119/1.2201854
3. Hipschman, R.: Exploratorium Cookbook, vol. II, Exploratorium, chap 63—Resonator (2002). https://www.exploratorium.edu/snacks/resonator

4. Laura, P.A.A., Pombo, J.L., Susemihl, E.A.: A note on the vibrations of a clamped-free beam with a mass at the free end. J. Sound Vib. **37**(2), 161–168 (1974). https://doi.org/10.1016/S0022-460X(74)80325-1
5. de Rosa, M.A., Franciosi, C., Maurizi, M.J.: On the dynamic behaviour of slender beams with elastic ends carrying a concentrated mass. Comput. Struct. **58**(6), 1145–1159 (1996). https://doi.org/10.1016/0045-7949(95)00199-9
6. Bigdeli, Y., Kim, D.: Damping effects of the passive control devices on structural vibration control: TMD, TLC and TLCD for varying total masses. KSCE J. Civ. Eng. **20**(1), 301–308 (2016). https://doi.org/10.1007/s12205-015-0365-5
7. Varanis, M., Silva, A., Mereles, A., Pederiva, R.: MEMS accelerometers for mechanical vibrations analysis: a comprehensive review with applications. J. Braz. Soc. Mech. Sci. Eng. **40**, 527 (2018). https://doi.org/10.1007/s40430-018-1445-5
8. Zalka, K.A.: A simplified method for calculation of the natural frequencies of wall-frame buildings. Eng. Struct. **23**(12), 1544–1555 (2001). https://doi.org/10.1016/S0141-0296(01)00053-0
9. Piccardo, G., Tubino, F., Luongo, A.: Equivalent Timoshenko linear beam model for the static and dynamic analysis of tower buildings. Appl. Math. Model **71**, 77–95 (2019). https://doi.org/10.1016/j.apm.2019.02.005
10. Cuvalci, O., Ertas, A.: Pendulum as vibration absorber for flexible structures: experiments and theory. J. Vib. Acoust. **118**(4), 558–566 (1996). https://doi.org/10.1115/1.2888335

Chapter 15
Guizzo Xp: A Robotic Toolkit for STEM Education and Raising Awareness of Aquatic Environment Protection

Daniele Costa, David Scaradozzi, Laura Screpanti, and Massimo Callegari

Abstract The necessity to prepare primary students with twenty-first-century skills through STEM-related teaching has considerably grown in the last two decades. Trainings, where robots are built from kits, programmed and tested, are a modern form of interdisciplinary education dedicated to children and youths. Classes in robotics are then expected to strongly impact on the development of mathematical literacy, scientific-technical information, and social competencies. To address this objective, the authors of this paper have developed a robotic toolkit which can be employed through a series of curricula as a case-study, design platform and test bench. Notably, the robot presented in this work is shaped like a fish and embodies all the motion functionalities of its biological counterpart, meaning the capability to swim in the aquatic environment and perceive its status. This paper also focuses on the group activities followed by primary students to learn the fundamentals of physics and machine mechanics before practicing with the mechatronic components of the toolkit. Finally, the bio-inspired architecture and the presence of the water element are aimed to raise awareness of aquatic environment protection in the younger generations.

Keywords STEM education · Robotic toolkit · Aquatic environment

---

D. Costa (✉) · M. Callegari
DIISM, Università Politecnica Delle Marche, Ancona, Italy
e-mail: d.costa@univpm.it

M. Callegari
e-mail: m.callegari@univpm.it

D. Scaradozzi · L. Screpanti
DII, Università Politecnica Delle Marche, Ancona, Italy
e-mail: d.scaradozzi@univpm.it

L. Screpanti
e-mail: l.screpanti@univpm.it

J. C. García Prada et al. (eds.), *Trends in Educational Activity in the Field of Mechanism and Machine Theory (2018–2022)*, Mechanisms and Machine Science 128,
https://doi.org/10.1007/978-3-031-25730-8_15

15.1 Introduction

Nowadays, engineering education pertains not only to undergraduate but also to pre-college education from an early age. Engineering disciplines and methods promote the development of many skills and support learning in STEM (Science, Technology, Engineering, Mathematics) subjects [1–3]. Thanks to the constructivist pedagogical theory and the pioneering work of Seymour Papert, robotics has entered education to boost children's big ideas through making, construction, programming and deploying a range of toolkits and machines [4]. Many educational robots have been developed over the years to support such activities [1, 2], and the way learners interact with the robot can help distinguish between different application fields, as described in [2, 5].

Usually, educational robots are focused on the exploration of activities for terrestrial or aerial environments. At the same time, only a few of them allow the exploration of the marine environment in pre-college education [2, 3, 5–9]. Besides, even if robots resembling animals or humans seem to be the most inclusive and engaging robotic tools [10, 11], biomimetic educational robots are not common, especially considering the aquatic environment. Although uncommon, some examples of biomimetic marine robots intended for educational purposes can be traced in the literature. In [12] Phamduy et al. propose a robotic fish as a tool for science education where users can remotely control the robot or watch it navigate autonomously. However, even if it is shaped like a fish and used for educational purposes, the robot seems to be remotely operated by children rather than programmed, like the educational robots used within the constructionist movement [1, 2, 4, 5]. Moreover, to the best of the authors' knowledge, no educational path is available to help schools integrate such tools.

The present paper illustrates the design of a fish-like robotic toolkit, "Guizzo Xp", whose aim is to engage students in meaningful learning about robotics by robotics and with robotics for formal and informal education in various learning environments. The paper is organized as follows: Sect. 15.2 illustrates the robot design, while Sect. 15.3 describes an educational path, which trains teachers about robotics, 3D printing and marine robotics, thus preparing them to integrate Guizzo Xp into their usual educational practice. Conclusions are discussed in the final Section.

15.2 Robot Design and Manufacturing

The robotic toolkit presented in this paper is shaped like a fish and is capable of swimming like its biological counterpart. The robot architecture consists of a rigid forebody, and an oscillating tail hinged to the prow section through a revolute joint. The structure has been designed to be easily assembled, inspected, and maintained by primary students supervised by untrained school personnel, following simple guidelines in the user manual. The robot can swim at constant depth by exploiting the tail thruster, which allows it to cruise in a straight path and perform turn maneuvers

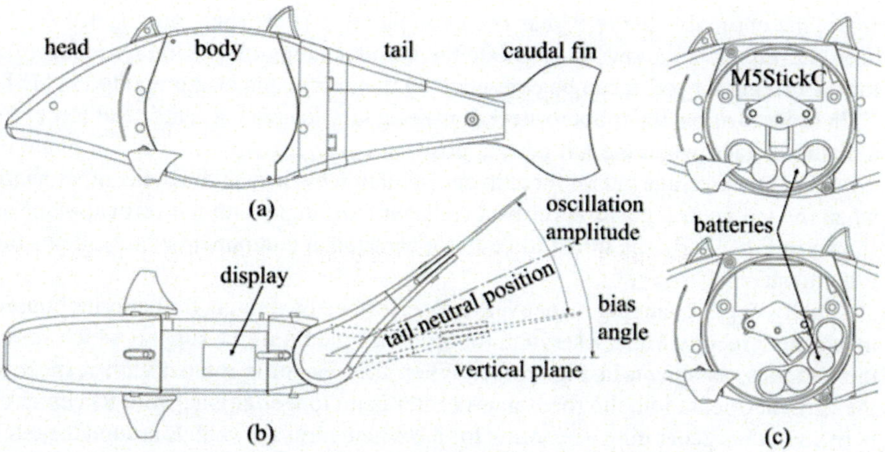

Fig. 15.1 **a** Robot blueprint; **b** steering configuration; **c** moving-ballast system

as well; moreover, a moving-ballast system installed in the forebody allows the robot to change its aptitude and perform climb and descent maneuvers. Both systems are detailed below. Figure 15.1a shows the blueprint of the robotic toolkit presented in this paper.

As stated before, the main propulsive system consists of the oscillating tail and the caudal fin it ends with. According to the nomenclature commonly adopted in the literature, the robot propels itself exploiting ostraciiform locomotion. Here, the rigid tail performs harmonic oscillations, generating the thrust necessary to balance the drag force and swim at constant velocity. Therefore, in order to size the tail thruster, the following equation can be written to balance the propulsive and resistance forces:

$$\frac{1}{2}\rho U^2 A_f c_D = \frac{1}{2}\rho U^2 c_T S_{fin} \tag{15.1}$$

where ρ is the water density, U is the cruising speed, A_f is the robot frontal area, c_D is the corresponding drag coefficient, whereas S_{fin} and c_T are the fin surface and average thrust coefficient [13]. The latter can be expressed in terms of the Strouhal number, a dimensionless parameter commonly used in oscillating-flow phenomena:

$$St = 2c \sin\theta_0 f/U \quad c_T = kSt^2 \tag{15.2}$$

where c, θ_0 and f are respectively the fin chord length, oscillation amplitude and frequency, whereas k is a dimensionless constant computed by the authors in previous works [14]. By replacing the Strouhal number expression in the thrust coefficient and the latter in (15.1), the following sizing equation can be written:

$$U^2 = f^2 4c^2 S_{fin} k/\left(A_f c_D\right) \sin\theta_0{}^2 \tag{15.3}$$

where the optimal value of θ_0 has been computed in [13]. Then, once speed U has been set, Eq. (15.3) allows to calculate the corresponding oscillation frequency. As for the driving torque, it can be computed by the expression also presented in [13].

In order to allow the robot to steer, it is necessary to create a centripetal force. As a matter of fact, when the tail neutral position coincides with the vertical plane of symmetry, the average lateral force in an oscillation cycle is null. On the other hand, when the tail neutral position forms a constant bias angle with the vertical plane as shown in Fig. 15.1b, the thrust force has a centripetal component which allows the robot to steer.

Finally, depth-changing maneuvers are performed by moving the batteries housed inside the forebody. Here, a servomotor drives a pendulum-like support which stores the batteries, as shown in Fig. 15.1c. When the system is rotated with respect to the vertical orientation, the robot mass center is no longer aligned with its center of buoyancy, thus generating a restoring torque which turns the vehicle around the pitch axis. Then, the simultaneous action of the tail thruster allows the robot to surface or dive.

The forebody section comprises two parts: the head, a hollowed thin shell shaped to resemble a carp, and the body, a waterproof compartment which houses the control and energy systems, together with the moving-ballast device discussed above. The watertight connection between the body and its cap is sealed by an O-ring housed inside the radial groove carved inside the cap. Both control and energy systems can be accessed remotely when the body is closed: this feature has been adopted to prevent the students from opening the toolkit, which could result in water leakages in case of incorrect assembly. The batteries can be recharged by placing the robot over its wireless charger, whereas the controller can be programmed by establishing a Wi-Fi connection with a computer or tablet running the development environment application.

Figure 15.2 shows the complete assembly of the robot, which is 350 mm long whereas the overall mass is 550 g. All its components have been 3D-printed by means of high-precision stereolithography. The main thruster is driven by a sealed servomotor embodied in the tail, which is manufactured as a thin and hollowed shell. Particularly, the servo motor, case and gear are fixed to the tail, whereas the output shaft and horn are bolted to the forebody. In this way, the driving torque is transmitted to the tail by means of a direct drive as in Fig. 15.1. Both the head and tail sections are flooded: thus, the only positive component is the sealed body, which produces enough buoyancy to prevent the robot from sinking. The chosen servomotor is capable of driving the tail and propelling the fish up to one body length per second, meaning about 300 mm/s, while the turnabout radius at full speed is two body lengths. Finally, the moving-ballast device is capable of tilting the robot 30 degrees about the pitch axis.

Several safety measures have been adopted to prevent primary students from harming themselves during the workshops. First and foremost, the toolkit has undergone a Risk Assessment carried on by a licensed company according to the current standards and has been evaluated to meet high safety and environment protection requirements, worthy to be sold as a product bearing the CE marking.

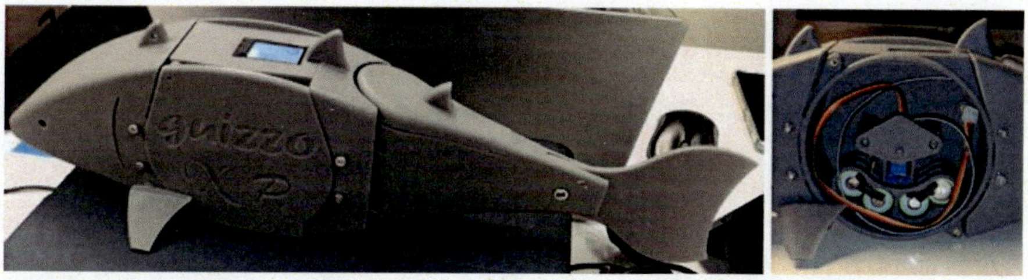

Fig. 15.2 Guizzo Xp: assembly and moving-ballast system

15.3 Workshop Activities

The following lessons and workshops have been conceived as educational and training material for schoolteachers with little or no experience in physics, programming, and robotics. Their content is an open-source resource that can be exploited to prepare classes and workshops for primary school students. The teachers are welcome to use any toolkit to organize the practical activities; however, the robotic fish presented in Sect. 15.2 has been specifically designed as the optimal tool for the lessons, since it embodies both the hardware and software presented in the course.

Practically speaking, the lessons plan can be considered an educational path course that can be further divided in two parts, as shown in Fig. 15.3 [15]. The first part focuses on basic concepts that are explored through curricular activities dedicated to one of the mechatronic components of the robotic fish, such as sensors and actuators. The second and more advanced course part is focused on the robot as a system and allows the student to program complex missions in an aquatic environment.

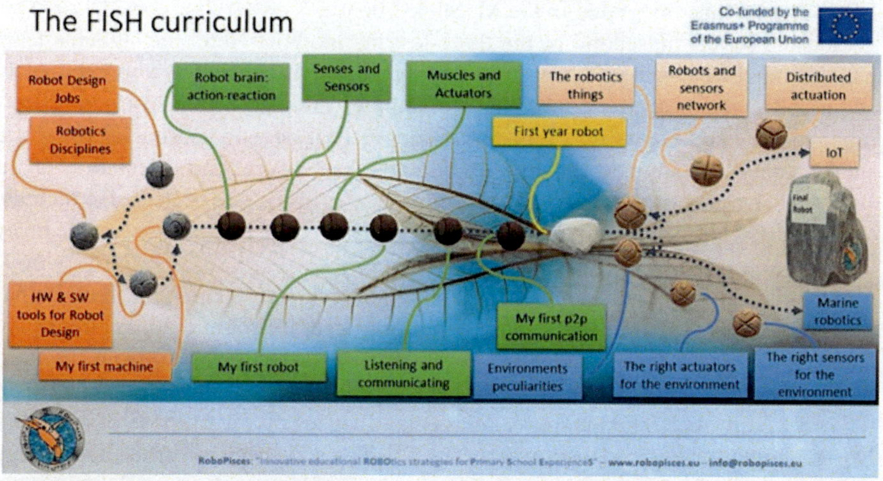

Fig. 15.3 The FISH curriculum [15]

15.3.1 Basic Course Activities

The course starts with a lesson which points out the differences between a robot and a machine: here, the focus of the class is autonomy, meaning how a robotic system perceives its surroundings, elaborates information, and implement suitable actions which involve the actuators. These concepts, supported by the concepts of inputs and outputs, are the backbone of the entire course and will be widely used throughout the different classes. As for the practical part, the students are invited to provide examples of machines and robots from their daily experiences.

The second lesson is focused on the roboticist job, particularly on the background skills necessary to design, manufacture, program, maintain and finally dismiss a robot to reduce its impact on the environment. In the practical sessions, the students learn how individual skills interact when a new robot is designed: individual roles, such as information and mechanical engineers, are assigned and exchanged during the course, which spans the life cycle of the system, from scratch to dismissal.

In the third class, the students start a three-lesson module focused on the functional blocks introduced at the beginning of the course: sensors, actuators, and brain, starting with the latter. Here, they are also introduced to the controller chosen for the toolkit presented in this paper, a M5StickC core element produced by M5Stack. Particularly, this class focuses on the hands-on activity, where the students learn the basics of coding by exploiting the board Integrated Development Environment (IDE) based on Blockly, an open-source software released by Google. The projects are then created by means of a visual programming editor running on Windows or Android web browsers, allowing the students to drag and drop the code blocks from the toolbox and then arrange them in the workspace area. Once created, the program can be easily uploaded in the controller and run to test its functionalities. Several tasks and exercises are thus assigned to the students to practice coding, focusing on loops, LEDs and buttons integrated in the M5StickC board.

Sensors cover the second of the three-lessons module. Here, the students learn the difference between exteroceptive and proprioceptive sensors by solving quizzes about self and environment awareness. Examples are also provided to compare a robot sensory equipment with human vision, touch, taste, and hearing. In the workshop activity, the students develop a code which measures the orientation of the M5StickC by means of its Inertial Measuring Unit (IMU). The project also includes the visualization of the aptitude by printing an arrow-based bubble level on the board display.

The introductory course ends with a lesson dedicated to actuators. Here, human muscles are compared to the motors commonly used in robotic applications. The students also learn that muscles contract when the order is passed from the brain through the nervous system; in the same way, motors need an output signal from the robot controller. The theoretical class also focuses on the purpose of actuators, like manipulation and propulsion: just like muscles power human arms, hands, legs, and feet, similarly, motors drive robotic arms, end effectors, wheels and thrusters. In the practical class, the students learn to control the servomotors installed inside the robotic fish by developing codes that they will use later to program missions.

15.3.2 Advanced Course Activities

The advanced course starts with a theoretical lesson on the aquatic environment and the importance of the water element for life sustainability in our planet. The students also learn the seven principles of ocean literacy, such as its significant influence on weather and climate, how it supports a great diversity of life and ecosystems, and how the ocean and human beings are inextricably interconnected. The class also focuses on the treasures and mysteries hidden in the sea, like the relics of ancient civilizations buried in archeological sites. In the following workshop, the students are introduced to the robotic fish: here, they learn its features and capabilities by practicing with the controller, sensors, ballast, tail and power unit. These preliminary activities are performed safely, on a workbench, to prepare for the upcoming "wet" operations.

In the second advanced class, the students are introduced to fluid statics: here, they learn Archimedes' principle and the conditions under which bodies are at rest in stable equilibrium when immersed in a fluid. The theoretical lesson also explains how pressure is related to depth. In the practical session, the students implement the concepts of buoyancy and aptitude to balance the robotic fish: they start by attaching known weights to the head and tail sections, until the back of the robot reaches the water free surface. Then, the students drive the moving-ballast system until the bubble level printed on the controller display, which is wrapped in resin on the fish body as shown in Fig. 15.2, indicates that the robot is balanced at zero-degrees pitch aptitude.

The third advanced class further expands the study of fluid mechanics by presenting the concept of dynamic equilibrium. Here, the students learn that motion is generated when the propulsive thrust balances the drag force exerted by water. In the following workshop, the tail is driven by running the code developed in the basic course. Thereafter, the students are asked to program the robot to navigate the path indicated by a series of buoys floating inside a pool, as shown in Fig. 15.4. To succeed in the assigned task, the students must figure out the beat cycles necessary to reach the buoys; then, they set the tail bias angle necessary to turn around them.

The advanced course ends with a final workshop: here, the students further practice with the robotic fish by programming a complete circle while surfacing the pool. To fulfil the task, they can exploit the controller buttons, which can be pressed remotely when the body is closed by approaching a magnet to the dedicated area.

Fig. 15.4 Advanced workshop activities in Rhodes, summer 2022

15.4 Conclusions

The importance of STEM-related teaching has considerably grown in the last twenty years to help students develop 21st century skills. Curricula, where robots are built from kits, programmed, and operated, are a modern form of interdisciplinary education dedicated to children and youth. To this end, the authors of this paper have developed a robotic toolkit that can be used as an educational tool to help students making practical experiences, thus implementing the theoretical concepts learned during the classes. The robot presented in this work is shaped like a fish and embodies all the motion functionalities of its biological counterpart. The bio-inspired architecture and the presence of the water element are expected to increase students' engagement in school education, promote their interest in STEM and Blue careers and raise the awareness of aquatic environment protection in the younger generations.

Further educational activities have been proposed to primary schools in addition to the course presented in Sect. 15.3. For example, students are encouraged to customize the robotic fish by drawing fins and appendices which can be later manufactured by commercial and inexpensive 3D printers to replace those provided within the toolkit. Additional options will be also available as soon as a camera will be added to the fish head to introduce the students to artificial vision, and wireless communication implemented in the robot controller to promote schooling and IoT activities. Lastly, Guizzo can be also exploited as an educational tool for secondary schools and universities as well to practice with advanced control techniques and multibody simulations.

In conclusion, the authors acknowledge the enthusiastic support of Manuele Silvestrini and Giacomo Fiara, bachelor students at Università Politecnica delle Marche and co-founders of ANcybernetics s.r.l, the university spin off created at UnivPM together with D.C. and D.S., to develop educational tools and promote environment protection.

References

1. Anwar, S., Bascou, N. A., Menekse, M., & Kardgar, A.: A systematic review of studies on educational robotics. *Journal of Pre-College Engineering Education Research (J-PEER)*, *9*(2), 2 (2019).
2. Scaradozzi, D., Screpanti, L., & Cesaretti, L.: Towards a definition of educational robotics: a classification of tools, experiences and assessments. In *Smart learning with educational robotics* (pp. 63–92). Springer, Cham (2019).
3. Scaradozzi, D., Cesaretti, L., Screpanti, L., Costa, D., Zingaretti, S., & Valzano, M.: Innovative tools for teaching marine robotics, IoT and control strategies since the primary school. In *Smart learning with educational robotics* (pp. 199–227). Springer, Cham (2019).
4. Blikstein, P., Kabayadondo, Z., Martin, A., Fields, D.: An assessment instrument of technological literacies in makerspaces and FabLabs. J. Eng. Educ. **106**(1), 149–175 (2017)
5. Siciliano, B., & Khatib, O.: Robotics and the Handbook. In *Springer Handbook of Robotics* (pp. 1–6). Springer, Cham (2016).
6. Bampasidis, G., Piperidis, D., Papakonstantinou, V.C., Stathopoulos, D., Troumpetari, C., Poutos, P.: Hydrobots, an Underwater Robotics STEM Project: Introduction of Engineering Design Process in Secondary Education. Advances in Engineering Education **9**(1), 1 (2021)
7. Ziaeefard, S., & Mahmoudian, N.: Marine robotics: An effective interdisciplinary approach to promote stem education. In *International Conference on Robotics and Education RiE 2017* (pp. 154–165). Springer, Cham (2017).
8. Groark, L., Cooper, K., & Davidson, D.: The SeaPerch evolution: from grassroots to a global community. In *2019 IEEE Integrated STEM Education Conference (ISEC)* (pp. 274–280). IEEE (2019).
9. Zande, J. M.: MATE ROV Competitions: Providing Pathways to the Ocean STEM Workforce. In *2011 ASEE Annual Conference & Exposition,* pp. 22–1043 (2011).
10. Toh, L.P.E., Causo, A., Tzuo, P.W., Chen, I.M., Yeo, S.H.: A review on the use of robots in education and young children. J. Educ. Technol. Soc. **19**(2), 148–163 (2016)
11. Pedersen, B. K. M. K., Weigelin, B. C., Larsen, J. C., & Nielsen, J.: Using educational robotics to foster girls' interest in STEM: A systematic review. In *2021 30th IEEE International Conference on Robot & Human Interactive Communication (RO-MAN)* (pp. 865–872). IEEE (2021).
12. Phamduy, P., Milne, C., Leou, M., Porfiri, M.: Interactive Robotic Fish: A Tool for Informal Science Learning and Environmental Awareness". IEEE Robot. Autom. Mag. **22**(4), 90–95 (2015). https://doi.org/10.1109/MRA.2015.2443991
13. Costa, D., Palmieri, G., Palpacelli, C., Callegari, M., and Scaradozzi, D.: Design of a bio-inspired underwater vehicle. In 12th IEEE/ASME International Conference on Mechatronic and Embedded Systems and Applications, Auckland, pp. 1–6, 2016.
14. Costa, D., Palmieri, G., Callegari, M., and Scaradozzi, D.: Multi-body Analysis of a Bio-inspired Underwater Robot. Mechanisms and Machine Science, Vol. 68, 2019.
15. RoboPishes website: https://www.robopisces.eu/

Chapter 16
Robotic Systems as a Part of AI Fundamentals Course at ITS Academy Foundation for New Life Technologies n.a. Alessandro Volta in Trieste, Italy

Giulia Zanin, Francesco Guzzi, Roberto Pugliese, Laura Cerni, Andrey Vukolov, Andrea Lorenzon, and Marco De Simone

Abstract The healthcare sector has always been the crucible for a high level of technology, as its use has a direct impact on life expectancy and the probability of recovery from a disease. In Italy the *ITS Academy Foundation for New Life Technologies Alessandro Volta* created AI-dedicated programs to educate skilled technicians in the development, setup, testing and service of complex software and hardware solutions for healthcare. The newest tools and methods make high use of Artificial Intelligence and robotics to help sanitary personnel in the diagnosis and treatment of diseases. In this paper, we describe the introduction of an AI and robotics class within the 2-year post-high-school program at the ITS Volta institute. The course gives students the ability to understand and manipulate the key components of complex software solutions, such as the one embedding robot control systems or AI algorithms.

G. Zanin · L. Cerni
Fondazione Istituto Tecnico Superiore per le Nuove Tecnologie della Vita Alessandro Volta, Basovizza, Italy
e-mail: info@itsvolta.it
URL: https://www.itsvolta.it/

F. Guzzi · R. Pugliese · A. Vukolov
Elettra Sincrotrone Trieste S.C.p.A., Basovizza, Italy
e-mail: info@elettra.eu
URL: https://elettra.eu

A. Lorenzon · M. De Simone
Central European Research Infrastructure Consortium (CERIC-ERIC), Strada Statale 14 - 163.5 km, AREA Science Park, 34149 Basovizza, Trieste, Italy
e-mail: info@ceric-eric.eu

A. Vukolov (✉)
IFToMM Permanent Commission for Education, Zakopane, Poland
e-mail: andrey.vukolov@elettra.eu
URL: https://iftomm-world.org/commissions/permanent-commission-for-education/

J. C. García Prada et al. (eds.), *Trends in Educational Activity in the Field of Mechanism and Machine Theory (2018–2022)*, Mechanisms and Machine Science 128,
https://doi.org/10.1007/978-3-031-25730-8_16

Keywords Robotics · Technical education · AI systems · ROS · TTS · RASA · Computer vision

16.1 Introduction

The ITS (Technological Institute of Higher Education) Academy Foundation for New Life Technologies Alessandro Volta, founded in 2014 in Trieste, Italy, is one of the over 100 Italian ITS foundations, renamed "ITS Academy foundations" by Italian law decree 99/2022, offering highly specialised 2-year courses corresponding to the V level of the European Qualification Framework [2]. ITS foundations were established in 2008 with the aim of filling the gap existing in Italy between the technological skills required by industries and those available on the labour market. They represent the first example of professional tertiary education in Italy, implementing a job-oriented educational approach (at least 60% of the teachers come from industries, lessons are in large part practical and at least 35% of the course is devoted to internships) and operating within six Technological Areas considered strategic for national economic development and competitiveness. In this context, the Academy Foundation offers three two-year courses funded by the European Social Fund Regional Operational Programme, managed by Autonomous Region Friuli Venezia Giulia. The Foundation's premises, located in AREA Science and Technology Park,[1] are characterised by state-of-the-art laboratories, in close vicinity with innovative start-ups and research centres. The laboratories include Lab3 (Laboratory for Advanced Technology in Healthcare Repair Training and Education), the first example in Europe of a laboratory entirely dedicated to the training of biomedical equipment technicians. Lab3 is designed to accurately reproduce a hospital environment, more precisely an operating theatre and two rooms for X-rays and Computer Tomography scans, allowing realistic simulations of different situations where the maintenance of machinery or the realisation of devices in compliance with hospital standards and protocols are needed. Lab3 also includes the bioinformatics laboratory, dedicated to the design, development, maintenance and managing of biomedical informatics systems, such as medical records, RIS (Radiology Information System), PACS (Picture Archiving and Communication System), LIS (Laboratory Information System). The newest addition to the premises is Lab3D, dedicated to Virtual and Augmented Reality, 3D Printing and Robotics. Lab3D was created in 2019 to boost the innovative component of the 2-year course "Engineering Technician for the Development and Management of Biomedical IT Solutions", which is the object of our case study. It provides all-round knowledge of ICT in the biomedical sector and is redesigned yearly in collaboration with partner companies to make sure that the learning modules respond to the employers' needs and are up to date with the

---

[1] https://www.areasciencepark.it/.

ongoing innovations. This course includes a module named "Fundamentals of Artificial Intelligence" which implements all the technologies and aspects described in the paper.

16.1.1 The Course

The 2000-h course curriculum provides all-round knowledge of ICT in the biomedical sector and is redesigned yearly in collaboration with partner companies to make sure that the learning modules respond to the employers' needs and are up to date with the ongoing innovations in hardware and software engineering. The course includes program for two years of tuition. The **first year** includes the following groups of modules and topics:

- Fundamentals of IT and Mathematics (60 h);
- Language, Communication and Relations (62 h);
- Science and Technology (101 h);
- Law and Economics (25 h);
- Organization and Management (35 h);
- Management, planning and quality control (61 h);
- Management of Healthcare Services (30 h);
- Fundamentals of Information Technology (231 h);
- Complements of Information Technology (200 h);
- Privacy and data security (60 h).

The mechanical engineering part is included into the program of the **second year** of tuition. This could be considered as response on the contemporary challenges [18] connected to infiltration of the robotics into the society. The technician should be ready to deal with existing robotic solutions, he also should be able to realize the development principles of modern robotics to understand the working issues and problems, to recognize and fix them. The full modular structure of the second year of the course is presented below:

- Fundamentals of Python Programming (30 h)
- Elements of Statistics for Artificial Intelligence (10 h)
- Fundamentals of 3D Programming (20 h)
- Elements of Medical IT I
  - Standard biomedical appliances: DICOM, HL7 and IHE initiative (30 h)
  - Software medical devices, apps and mHealth I (50 h)
  - Fundamentals of information systems maintenance (installation and update) (40 h)

- Elements of Medical IT II
  - Templates for clinical and nursing records and social welfare and healthcare data folders (40 h)
  - Models of hospital and extra-hospital processes (30 h)
  - Software medical devices, apps and mHealth II (50 h)
  - Fundamentals of IoHT-Cloud-Azure-AWS (50 h)
  - High Performance Computing platforms (30 h)
- Augmented Reality, Virtual Reality and AI
  - Elements of Augmented Reality, Virtual Reality and Mixed Reality (40 h)
  - Fundamentals of Artificial Intelligence (60 h)

16.1.2 "Fundamentals of Artificial Intelligence": The Learning Module

The learning module Fundamentals of Artificial Intelligence was introduced in the general curriculum of the course, with the aim of giving the students an understanding of AI and how it is currently applied to multiple fields, including the biomedical sector. In the first part, students learn about the impact of exponential technologies on society and acquire the basic terminology and fundamental concepts of Artificial Intelligence, Machine Learning and Data Science.

After being given a solid foundation of programming languages (Python primarily) and software engineering, students are guided towards the comprehension of most common data analysis, data visualisation and data science algorithms and tools. Core concepts and practical real-life applications and use cases are preferred against a more theoretical, academic approach, with the aim of giving the students the skills required to assess robotics and data science problems, choose existing tools and algorithms, optimise iterations of their use, evaluate their performance, and present results in elegant, effective formats.

Then, they learn how to use Artificial Intelligence, Machine Learning and Data Science in a company and how to organise a project. The practical part of the module focuses on Robotic Process Automation and how it can be used in the company to automate repetitive tasks by freeing up human resources that can be used for higher-level activities; it also focuses on what Voice User Interfaces (VUIs) and Chatbots are, how they work and what technologies they are based on. The robotic solutions are also used to give a practical explanation on the applications of mechanical engineering in the real life, so the course involves both the industrial and commercial robots, and prototypes or prototyping platforms. These activities are described below.

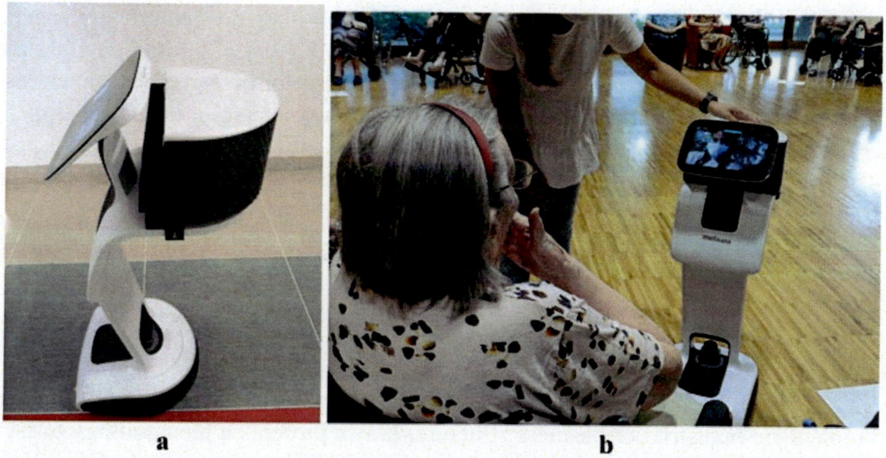

Fig. 16.1 Medibot robot with mounted Smart Delivery compartment

16.1.3 Relations Between the Course and Innovative Robotic Solutions Used in ITS

During the year program, students had the opportunity of applying the skills and knowledge gained from the Fundamentals of Artificial Intelligence module to ITS 4.0,[2] a competitive programme funded by the Ministry of Education and addressed to all Italian ITS Foundations. In the programme, participants were asked to use the Design Thinking method to develop a project in collaboration with a company. A total of 76 projects participated, divided into five categories: Digital Services, Sustainability, Immersive Technologies, Smart Industry, Fashion and Design.

In the 2021–2022 programme, ITS Foundation Alessandro Volta presented Medibot, a home care robot based on Temi platform[3] (Fig. 16.1a), that supports the staff and guests of retirement/nursing homes in their daily needs.

The project was developed in collaboration with `VisionQub.It`,[4] an innovative start-up that develops cutting-edge systems and services for robotics and Artificial Intelligence solutions. The robot, in semi-autonomous mode, performs three functions: Smart Delivery (safe delivery of documents and drugs), Telepresence (video calls with family members and other external people) and Guide Me To (support to guests in orienting themselves in the structure). The Smart Delivery function required the 3D printing of a box with two compartments equipped with a smart locker which was done by `VisionQub.It`. In the final event of ITS 4.0 competition, Medibot ranked first in the Digital Services category. This robot was also used to demonstrate for the students, how the smart systems are interacting with the people that have no

[2] https://www.its40.it/wp/.

[3] https://www.robotemi.com/en/.

[4] https://visionqub.it/.

special skills to operate and maintain (Fig. 16.1b). It does not require any special knowledge, so it can be easily used as the platform for initial training. Within the course, Medibot acts successfully as a "primary attractor" for students' attention, orienting their eyes to more sophisticated solutions and technologies. Then the general module divides into the branches or submodules which are described below. The overall module lectures is given by ITS teachers with participation of external engineering practitioners from Elettra Sincrotrone Trieste.[5]

16.2 Artificial Intelligence Applications

This submodule explains how to identify the opportunities to apply AI to existing real problems in industrial or business structure. It also presents a methodology to face Machine Learning and Data Science projects as well as how to navigate the ethical and social discussions surrounding Artificial Intelligence. Artificial Intelligence here is shown as one of the most important technologies for the modern world. The robotics is also explained as an ideal context where AI can turn fundamental to build smart solutions with native language-compatible conversational interfaces. This submodule also gives the students a clue about the principles of high-level development of the most sophisticated industrial systems explained in the other submodules.

16.2.1 Computer Vision Submodule

Machine Learning and Computer Vision submodules aim at educating the students on the practical aspects of the most important computer vision solutions, making them competent in integrating off-the-shelf methods within their own programs, and they are designed to make an explanation manner suitable for undergraduates. Differently from the past, that field appears no more as a niche of computer science, as in many cases, it represents instead what triggers the interest of lots of students in such a career [15]. Therefore, a honest and clear approach is required to build upon the initial student interest. The whole submodule is subdivided into the sections: Machine Learning basics (the concept of model, data scrubbing, features, training, estimating a quantity, classifying an input, implementing pipelines in Tensorflow), Image processing basics, Common problem (programming an object detector), Example workflow for face recognition. The special part here is the usage of an embedded MyriadX/OAK-D [9] device installed on the real robot (see the detailed explanation below).

As it can be seen, the topics of the submodule are organised to prepare the students for Face Recognition, which is an interesting CV problem, not only due to its usage and effect on modern society, but also from the tuition point of view: indeed, it allows

[5] https://elettra.eu/.

to briefly cover many topics, such as Convolutional Neural Networks (CNN), image processing pipelines, objects detection, training methodologies, features generation, feature space, classification, database manipulation.

Each lesson of the submodule is taught similarly to a tutorial, where practising the implementation of the topic under tuition is the most important thing. The lecturer acts here as a true practitioner. Indeed, theoretical explanations are immediately followed by an exercise which can either be used to test comprehension or to provide proof and further explain the concepts at hand.

In many contexts, a series of simplified libraries are provided to the students [13, 15]. Here, instead, we prefer to use standard computer vision/image processing tools and libraries and a fully featured programming language, due to the high professional orienting nature of these courses. A good compromise is provided by the use of Jupyter Notebooks, that at the same time gives attention to Python coding and allows one to follow a "hands-on" approach while teaching new concepts. Thus each lesson in the submodule is built on an interactive reading of a Jupyter Notebook `ipynb` file, where pre-rendered webpage-like sections form the theoretical "material" for the lesson, and interactive code blocks constitute the playground environment. An essential part of the hands-on approach lies in presenting the software tools which can be used to solve a given problem, namely the software libraries to import, process and visualise data.

A short introduction to the fundamentals of image processing is required to proceed on CV topics. To do so, Numpy, `scikit-image`, OpenCV[6] and `matplotlib` are the preferred Python tools to teach and explore these concepts, especially if paired with the interactive GUI controls embedded in Jupyter. Machine Learning is introduced in its two fundamental functions: to provide models which can estimate a numerical quantity or to classify an input. The aforementioned blocks constitute the basics to analyse object detection and recognition, which is one of the glorified applications of narrow AI [5]. To introduce image-oriented convolution neural networks (CNN), a simple model trained on Fashion MNIST [1] is designed and tested. The characteristics of the dataset allow designing a state-of-the-art but simple CNN of a few different layers which can be trained in minutes on a modern consumer laptop. Object detection models can be trained similarly only if very simple, otherwise, complex Deep Learning pipelines are required. That is why within the course, only pre-trained models are explored and libraries to fine-tune a model on a custom dataset are only enunciated. The SSD and YOLO [4] object detectors are briefly presented and their OpenCV implementation is tested. With this machinery, a complex state-of-the-art Deep Learning pipeline for face recognition is then introduced, analysing how to implement each processing phase: face detection, face alignment, feature extraction, feature classification, output visualisation, and identity handling. This sequence is common to many other problems of CV. We follow the implementation of the system reported in [3, 6, 7, 10], where a pre-trained `resnet18` Dlib model [8] is employed to extract the features from each input image. The course ends with a hands-on activity on edge processing: the face recognition architecture

[6] https://opencv.org.

is implemented on an edge device called OAK-D,[7] which is a triple camera system paired with Intel/Movidius MyriadX [9] tensor accelerator for which the feature extraction network has been deployed. This kind of device is becoming pervasive as it allows small single-board PC mounted on robots to compute complex image processing tasks. Knowing how to implement an example deep learning pipeline on such a device is a highly desirable skill for an AI technician, so the submodule gives a point for achieving this skill.

16.2.2 Conversational AI Submodule

The Conversational AI submodule presents the design fundamentals of voice user interfaces and voice enabled applications. Different enabling technologies are presented and specific applications and case studies are implemented using RASA AI [14] that is an open source technology that the students can use on their own and in future deploy easily in the companies they will work in. The students learn the answers for the following questions:

- What are Voice User Interfaces (VUIs) and how do they work and what technologies are they based on?
- What are Chatbots, how are they related to VUIs and what technologies are they based on?
- How can VUI and Chatbot be introduced on high level into the development pipelines of the companies' products and activities?

At the end they are even asked to develop a simple chatbot, a FAQbot on their own in order to make their learning experience practical.

16.3 Autonomously Navigating Robotic Systems

The Medibot robot described above is part of a set of robots used in the course. It should be considered as the most complete solution where the most control tasks, programming aspects and internal data interaction are hidden from the end user. It is obviously good for the amplification of interest, but deeper explanation of the control principles requires more open systems allowing the lecturer to use intermediate data and signals during the lecture. To avoid limitations, two other robots are used. These robots are the prototypes developed by Elettra Scientific Computing Team. The first is 60-kg indoor industrial-grade transportation machine (Fig. 16.2a) manufactured

[7] https://www.luxonis.com/.

a b

Fig. 16.2 Eutronica Jobot industrial robot (**a**) and Rover Zero 3 prototyping platform (**b**)

by Eutronica S.R.L.[8] (Italy). The second one is Rover Robotics Zero 3[9] (Fig. 16.2b) provided by Rover Robotics (USA).

The robots are built on different chassis but they are using the same state-of-the-art super-automatic control system. This system includes vision-based docking controller [19], AI video processing solution based on Intel MyriadX chip, and autonomous navigation system. The navigation is implemented using Google Cartographer LIDAR processing engine allowing mapping and then navigating on actually every area feasible for LIDAR. The robot carries totally open source ROS 1[10] as a backend framework that exposes a relation graph easily understandable by the students [12]. The navigation system involves a specially developed global path planner library optimized for sparse maps and dynamic obstacles. This library is built over D-Star greedy informed algorithm of path planning [16] allowing the robot to rebuild the terrain maps for dynamically recognized obstacles with expensive initial planning and fast replan. ROS tools and programs [17] are used to demonstrate visualization, planning and mapping processes to the students with visible explanation and a possibility for the teacher let students feel the control stage between the frontend solutions and hardware. During the lectures dedicated to ROS, the students' involvement increases significantly when the ordinary explanation of the design principles is accompanied with visual representation. ROS tools allow to do it easily with real-

[8] https://www.eutronica.com/.

[9] https://roverrobotics.com/collections/all/products/copy-of-4wd-rover-zero-3.

[10] http://wiki.ros.org.

Fig. 16.3 Mapping process explanation using Jobot

time experience on observation: how the elements of the system change their states reacting to the controllable incoming data. For example, the mapping software is explained directly when the students are driving the robot over the building allowing LIDAR to collect the data for the statistical matching engine (Fig. 16.3).

From the other side, the web interface can be explained and considered as full fledged subsystem with own functions and interfaces. The special mission control language is developed and involved to allow the user to program the complex operations performed by the robot as a set of transactions directly from the user interface. Additionally, it allows the students also to understand the clue between high-level development and onboard systems of the robot. An explanation of the real robotic systems integrated with high-level solutions completes the course and shows the student the results of all the work done in the industry.

The prototyping platforms and real industrial robots are also used to give the detailed explanation about how the modern control frameworks work in terms of mechanical engineering. The tools already mentioned before allow the teacher to explain and to get the student in touch with the real mechanical actuation devices inside the robot, and the design patterns used to control them from the software. As an example, Fig. 16.4 presents the ROS node graph used in the course to explain data flow between the controller software components dedicated to the tasks served by actual devices. Using even a minimal experience with ROS structures, the teacher can now explain existence of the points where the "virtual" data circulating inside the control system are converted into the commands and signals triggering the activities which formulate the robot behaviour.

As a frontend solution gathering everything done through the development of the robot, the web application is used (Fig. 16.5). This web application allows end user to control the robot manually, and also to program, run and schedule missions. It creates for the student a full experience about how the frontend under-development

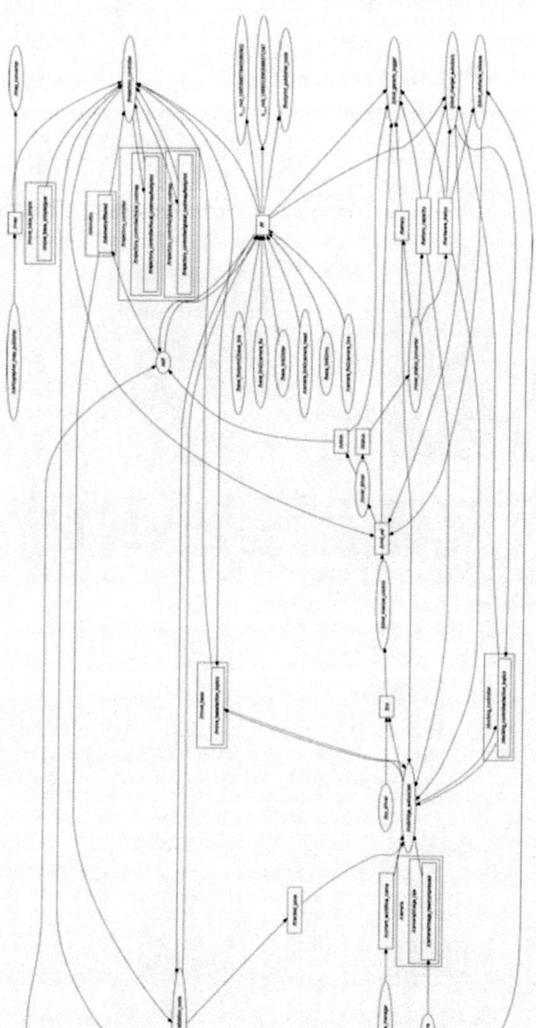

Fig. 16.4 Example of the node graph

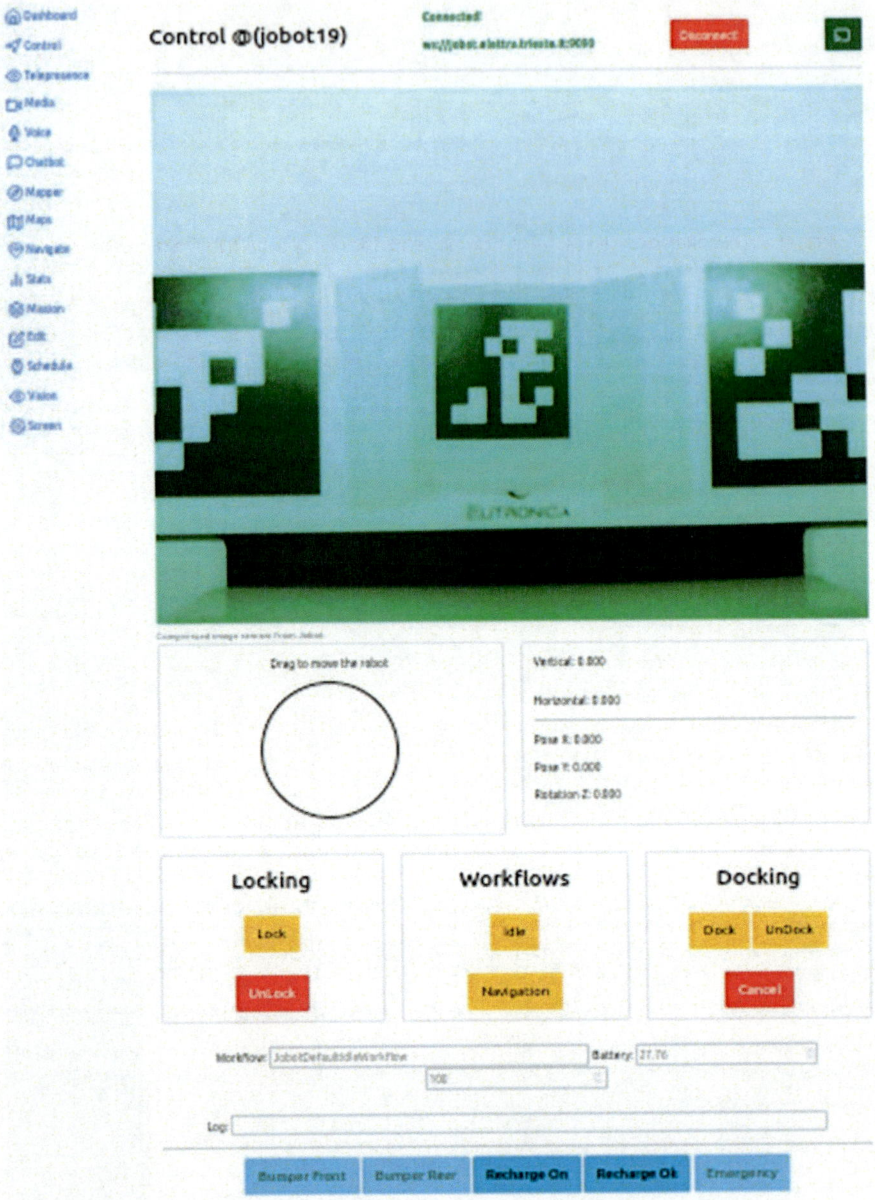

Fig. 16.5 System frontend web application

applications work and how they orchestrate the interaction between the other components. Also, the ROS integration used here is a great opportunity for the lecturer to explain the students how the interprocess communication frameworks work. The total openness provided by such a system significantly improves the quality of transferred knowledge [11, 20] because the understanding of the processes studies is now accompanied with the ability to control them and to discover how the high-level user commands influence the whole robotic solution.

16.4 Conclusion

In the presented paper we discussed the experience of implementation of the learning module dedicated to ICT, Robotics and Artificial Intelligence at ITS Academy Alessandro Volta. The described module with its submodules allowed to significantly improve the quality of professional orienting and skill development for undergraduate students. A possible path for introducing these technologies in the health system is also given in the course. The real robotic solutions developed by different vendors are used within the course to explain and introduce the links between all the stages of development. The students are allowed to touch the technologies directly and to feel the difference between the high-level frontend interfaces and underlying control system, sensors and software. The course also allows the lecturer to concentrate on explanation of the material itself because the involved "tech playgrounds" attracts the attention focus of the students, keeping them ready to think about the new information in the context of "the present moment" of their real lives. In fact, it implements the classic tuition techniques directly on the modern AI-oriented environment involving all their advantages with eased understanding provided by computers.

References

1. Bruna, J., Zaremba, W., Szlam, A., LeCun, Y.: Spectral networks and locally connected networks on graphs (2013). arXiv preprint arXiv:1312.6203
2. Council of European Union: Council recommendation of 22 May 2017 on the European Qualifications Framework for lifelong learning and repealing the recommendation of the European Parliament and of the Council of 23 April 2008 on the establishment of the European Qualifications Framework for lifelong learning. EUR-Lex. https://eur-lex.europa.eu/legal-content/EN/TXT/?uri=CELEX%3A32017H0615%2801%29&qid=1663960811535. Document 32017H0615(01)
3. De Bortoli, L., Guzzi, F., Marsi, S., Carrato, S., Ramponi, G.: A fast face recognition CNN obtained by distillation. In: Saponara, S., De Gloria, A. (eds.) Applications in Electronics Pervading Industry, Environment and Society, pp. 341–347. Springer International Publishing, Cham (2020)
4. Du, J.: Understanding of object detection based on CNN family and YOLO. J. Phys. Conf. Ser. **1004**, 012029 (2018). IOP Publishing
5. Goertzel, B., Pennachin, C.: Artificial General Intelligence, vol. 2. Springer (2007)

6. Guzzi, F., De Bortoli, L., Marsi, S., Carrato, S., Ramponi, G.: Distillation of a CNN for a high accuracy mobile face recognition system. In: 2019 42nd International Convention on Information and Communication Technology, Electronics and Microelectronics (MIPRO), pp. 989–994 (2019). https://doi.org/10.23919/MIPRO.2019.8756937
7. Guzzi, F., De Bortoli, L., Molina, R.S., Marsi, S., Carrato, S., Ramponi, G.: Distillation of an end-to-end oracle for face verification and recognition sensors. Sensors **20**(5) (2020). https://doi.org/10.3390/s20051369
8. King, D.E.: Dlib-ml: a machine learning toolkit. J. Mach. Learn. Res. **10**, 1755–1758 (2009)
9. Marantos, C., Karavalakis, N., Leon, V., Tsoutsouras, V., Pekmestzi, K., Soudris, D.: Efficient support vector machines implementation on Intel/Movidius Myriad 2. In: 2018 7th International Conference on Modern Circuits and Systems Technologies (MOCAST), pp. 1–4 (2018). https://doi.org/10.1109/MOCAST.2018.8376630
10. Marsi, S., De Bortoli, L., Guzzi, F., Bhattacharya, J., Cicala, F., Carrato, S., Canziani, A., Ramponi, G.: A face recognition system using off-the-shelf feature extractors and an ad-hoc classifier. In: Saponara, S., De Gloria, A. (eds.) Applications in Electronics Pervading Industry, Environment and Society, pp. 145–151. Springer International Publishing, Cham (2019)
11. Procurat, G., Strukova, A., Vukolov, A., Egorova, O.: Generated graphics and game development software in engineering education: perspectives and experience of usage. In: García-Prada, J., Castejón, C. (eds.) New Trends in Educational Activity in the Field of Mechanism and Machine Theory, pp. 123–130. Springer International Publishing, Cham (2019)
12. Quigley, M., Conley, K., Gerkey, B., Faust, J., Foote, T., Leibs, J., Wheeler, R., Ng, A.Y., et al.: ROS: an open-source robot operating system. In: ICRA Workshop on Open Source Software, vol. 3, p. 5. Kobe, Japan (2009)
13. Reimer, P., Branzan Albu, A., Tzanetakis, G.: Raydiance: A tangible interface for teaching computer vision. In: International Symposium on Visual Computing, pp. 259–269. Springer (2011)
14. Sharma, R.K., Joshi, M.: An analytical study and review of open source chatbot framework, rasa. Int. J. Eng. Res. **9**(06) (2020)
15. Sigut, J., Castro, M., Arnay, R., Sigut, M.: OpenCV basics: a mobile application to support the teaching of computer vision concepts. IEEE Trans. Educ. **63**(4), 328–335 (2020)
16. Stentz, A., et al.: The focussed D* algorithm for real-time replanning. IJCAI **95**, 1652–1659 (1995)
17. Vukolov, A., Baskara, zacharykratochvil: ElettraSciComp/witmotion_IMU_ros: Version 0.8.22 released (2022). https://doi.org/10.5281/zenodo.7015494
18. Vukolov, A., Egorova, O.: New challenges of 3D mechanisms models in TMM training. Int. Rev. Mech. Eng. (IREME) **10**(5) (2016). https://doi.org/10.15866/ireme.v10i5.9246. https://www.praiseworthyprize.org/jsm/index.php?journal=ireme&page=article&op=view&path%5B%5D=18997
19. Vukolov, A., Kourousias, G., Pugliese, R.: Flexible vision-based auto-docking control system for unmanned ground vehicles equipped with differential chassis. In: Müller, A., Brandstötter, M. (eds.) Advances in Service and Industrial Robotics, pp. 403–411. Springer International Publishing, Cham (2022)
20. Vukolov, A., Titov, A., Prokurat, G., Lapteva, M.: 3D application for modeling of involute gear manufacturing as the assistance solution for TMM training. In: Uhl, T. (ed.) Advances in Mechanism and Machine Science, pp. 813–823. Springer International Publishing, Cham (2019)

Chapter 17
The Potential of Education and Training in Additive Manufacturing

Álvaro Rodríguez-Prieto, Juan Claver, Jorge Ayllón, Amabel García-Domínguez, Ana María Camacho, and Miguel Ángel Sebastián

Abstract The final aim of education and training in additive manufacturing is to reach competences and skills in 3D printing, but its potential in different fields and scopes of application goes beyond it. This paper presents an overview of the competencies and skills that must be developed to train students in this field at different educational levels; but also presents a set of dimensions with potential application of additive manufacturing, which allows us to understand the need for comprehensive training in this field and its role in the era of digitization and education in STEAM subjects.

Keywords Additive manufacturing · 3D printing · Education · Training · Applications · STEAM subjects

17.1 Introduction

Additive manufacturing (AM), commonly known as 3D printing, has emerged in recent years not only as a technological alternative for the manufacture of certain parts and products, but also as a field of experimentation, that allows exploring new forms of design and production, giving solutions to past, present and future challenges.

The final aim of education and training in additive manufacturing is to reach competences and skills in 3D printing [1], but its potential in different fields and scopes of application goes beyond it [2]. Thanks to its extraordinary geometric freedom, additive manufacturing has not only made it possible to find optimized solutions to existing problems, but also to propose new ones that were previously unimaginable. Its synergies with other technologies, such as 3D modeling, parametric

Á. Rodríguez-Prieto · J. Claver · J. Ayllón · A. García-Domínguez · A. M. Camacho (✉) · M. Á. Sebastián
Universidad Nacional de Educación a Distancia (UNED), Juan del Rosal 12, 28040 Madrid, Spain
e-mail: amcamacho@ind.uned.es

J. C. García Prada et al. (eds.), *Trends in Educational Activity in the Field of Mechanism and Machine Theory (2018–2022)*, Mechanisms and Machine Science 128,
https://doi.org/10.1007/978-3-031-25730-8_17

design or optimization tools, are catalysts that further enhance the possibilities of additive technologies.

Moreover, these technologies have had great acceptance and interest from the society, which has been interested in their possibilities from very heterogeneous fields (productive, technological, social, cultural or educational), which is another proof of the importance of exploring all its dimensions.

It is possible to define some categories related to the development of competences in Additive Manufacturing in technical studies, namely:

- Category 1. Additive manufacturing technologies
- Category 2. Design for additive manufacturing
- Category 3. Materials for additive manufacturing
- Category 4. Data of the 3D printing process.
- Category 5. Post-processing operations
- Category 6. Regulatory and certification aspects.
- Category 7. Production management

Obviously, these competences must be adapted to the different educational levels and to the profile of the studies. This paper presents an overview of the competencies and skills that must be developed to train students in this field at different educational levels, but also presents a set of dimensions with potential application of additive manufacturing, which allows us to understand the need for comprehensive training in this field and its role in the era of digitization and education in STEAM subjects (science, technology, engineering, art and mathematics).

17.2 Training Programmes

17.2.1 Upper Level Training Cycles

The recent publication of Royal Decree 280/2021, of April 20, establishes the Specialization Course in "Additive Manufacturing for Higher Level Professional Training" and the basic aspects of the curriculum [3]. As indicated in the aforementioned Royal Decree, the general competence of this specialization course is to develop and manage additive manufacturing projects through the use of 3D printing, supervise or execute the assembly, maintenance and start-up of these projects. As well as making implementation decisions in the development of company products, considering criteria of quality, design, safety and environmental aspects.

17.2.2 Bachelor and Master

Focused on University studies, specific guidelines have not been defined yet in order to include additive manufacturing in subjects of regulated education curricula such as

Bachelor's and Master's degrees in Engineering Industrial. Thus, a propose of competences to be developed in Master studies is included in Table 17.1. Competences that are shared with undergraduate studies, are marked in bold. The Higher Technical School of Industrial Engineering (ETSI Industriales) of UNED already offers training in Additive Manufacturing through the Official Master's degree "Advanced Manufacturing Engineering" (Ingeniería Avanzada de Fabricación) with subjects such as "Additive manufacturing technologies" and "Supply chain management in Industry 4.0"; and shortly in the Master's degree "Connected Industry", with subjects "Additive Manufacturing in Connected Industry" and "Advanced Technologies of Manufacturing".

Within the offer of official studies, a promising sort of programs are the so-called Micro-degrees, being able to provide general or specific training depending on the design of the syllabus.

17.2.3 Longlife Training

In addition to official studies, the longlife training has an important role to harness the potential of additive manufacturing in different fields and scopes of application. There are different programs in the national context, but the Modular Program of Additive Manufacturing (Programa Modular/Máster en Fabricación Aditiva) of UNED [5] is a singular one as it is given under agreement with the Consejo Superior de Investigaciones Científicas—CSIC, with experts belonging to the Plataforma CSIC para el desarrollo de la Fabricación Aditiva FAB3D [6].

17.3 Methodological Aspects

In order to explore the potential of education and training in additive manufacturing, it is necessary to define the application, challenge or problem to be solved. Some examples of different dimensions with potential using AM are the following ones: (a) Manufacturing of complex geometries; (b) Achievement of advanced functional requirements; (c) Maintenance and reparation tasks; (d) New designs and prototyping; (e) Resource in teaching and learning; (f) Alternative to social problems; (g) Applications related to Heath; (h) Advantages from a sustainable approach.

Depending on the application, one or different dimensions with potential can apply. And depending on the case, a general or specific training a needed. A sketch that summarizes all the aspects involved is presented in Fig. 17.1. Hereafter some examples of applications where AM show its potential are presented. These are practical examples of opportunities of the 3D printing as disruptive technologies in the era of digitalization.

Table 17.1 Main professional competences and skills in education and training in Additive Manufacturing for Master programs[1]

Category	Description
1. Additive manufacturing technologies	• **Describe and differentiate the AM technologies, analyzing the general vision of process categories and raw materials according to the UNE-EN ISO 17296–2 standard** [4]
2. Design for additive manufacturing	• **Recognize dimensional precision in 3D printing technologies** • Build and apply the concept of topological optimization • Manage and transform a design for additive manufacturing (DfAM) • Employ reverse engineering concepts for additive manufacturing • **Skill in the use of soluble support filaments**
3. Materials for additive manufacturing	• Select and differentiate advanced 3D printing materials (polymers, metals, composites, loaded, special) • Handle special materials management applications in 3D printing
4. Data of the 3D printing process	• **Programming capacity of the process variables** • **Validate and control the use of any STL repair and lamination software** • Manage and transform the mesh of the piece according to needs • **Ability to calculate and program mounting tolerances between printed assemblies** • **Ability to recognize the glossary and terminology of 3D printing** • Ability to apply 3D printer calibration routines • Determine the cost estimate for a 3D printed part
5. Post-processing operations	• **Ability to remove supports in printed parts** • **Ability to clean printed parts** • **Ability to use electrical cutting, cleaning parts** • Control the use of paint coating technique • Control the use of welding technique • Control the use of the paste technique • Other post-processing
6. Regulatory and certification aspects	• **Being able to handle the reference regulations in additive manufacturing** • Acquire knowledge about the main inspection techniques of components obtained through additive manufacturing

(continued)

[1] Those competencies that are shared with undergraduate studies, are marked in bold.

Table 17.1 (continued)

Category	Description
7. Production management	• **Ability to compare additive manufacturing processes with other manufacturing methods in terms of production time, quality, cost and flexibility** • **Ability to take advantage of the potential of additive manufacturing for a more sustainable production** • Ability to integrate additive manufacturing into production systems, meeting the requirements of Industry X.0

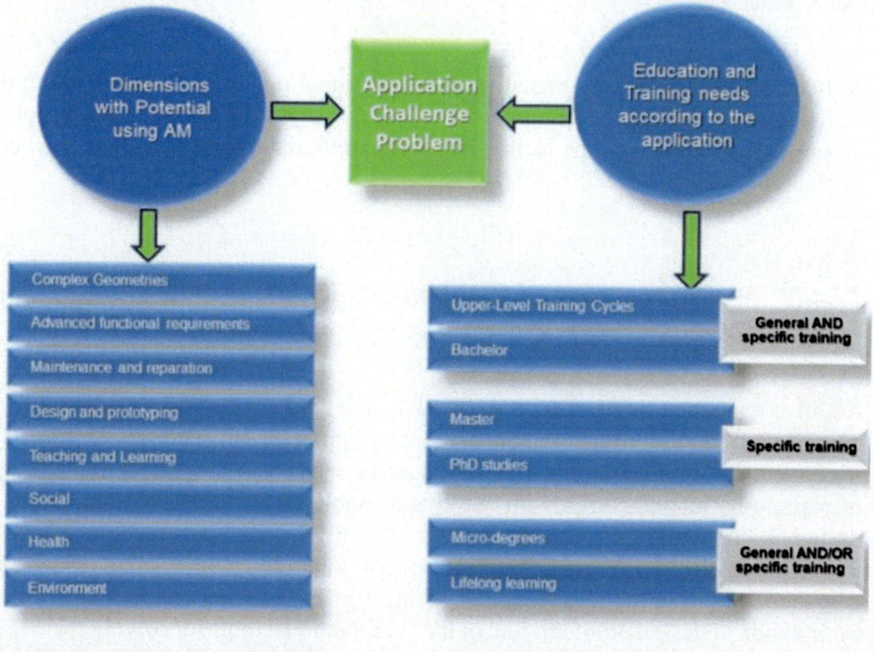

Fig. 17.1 Summary of the dimensions with potential using AM and different training programs depending on the application, challenge or problem to be approached

17.4 Examples of Application in Industrial Engineering

17.4.1 Manufacturing of Components as an Alternative to Non-additive Processes

Thanks to the important scientific-technical development of AM technologies in the last decade it is already possible to fabricate fully functional parts, and therefore,

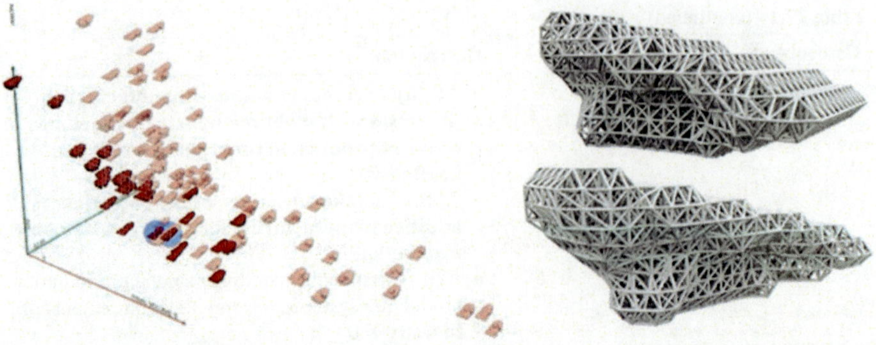

Fig. 17.2 Multi-objective optimization algorithm result for a shoe last design to be produced with additive technologies [11]

AM can be considered an alternative to conventional processes in cases where the high customization of the component prevents its mass production, among other examples. Depending on the technology to be used, the training should be general or specific.

17.4.2 Manufacturing of Components with Complex Geometries

Additive manufacturing is not constrained by the restrictions of conventional technologies and allows the manufacture of complex geometries that are very expensive to manufacture with other technologies. This represents a paradigm shift in the design of parts to be manufactured with additive technologies. Due to the complexity of design problems and the resulting geometries, it is essential to have specific software and algorithms (Fig. 17.2) to help the designer [7–9]. Therefore, specialized training is essential, both in the knowledge of additive manufacturing technologies for the formulation of design problems and of the computational tools for solving them [10].

17.4.3 Manufacturing of Lightweight Structures

One of the greatest potentials of additive technologies is the manufacture of lightweight parts, since it is possible to obtain geometrically complex parts with internal cavities. Lightened designs allow, on the one hand, to practically eliminate material waste in the manufacturing process, and on the other hand, to reduce energy consumption due to the use of these lightened components in use, for example, in the aeronautical sector [12].

17.4.4 Reparation of Parts of Widespread Use

The presence of additive manufacturing equipment in home environments, in small businesses and even in our children's schools, shows how close this technology is to society. This knowledge and interest in these technologies, combined with a significant increase in the possibilities of accessing low-cost equipment, pose scenarios that were previously difficult to imagine, in which a growing number of end users of certain products can act as designers and manufacturers of parts to replace damaged ones.

17.4.5 Maintenance Through Reverse Engineering

In some industries such as electricity generation or in sectors such as petrochemicals, in recent years difficulties have been encountered in replacing certain critical components that, with a complex geometry, together with discontinuous production over time, have given rise to an obsolescence problem. AM has turned out to be a solution to this problem, as these parts and their designs are very difficult to obtain due to the lack of design information, such as manufacturing drawings or bills of materials. In other cases, the manufacturer does not provide manufacturing data [13]. In these cases, a solution is the application of a reverse engineering process to establish design and manufacturing requirements.

17.4.6 Maintenance Tasks Through DED Techniques

Another field with potential of AM technologies, especially in the field of metal components, is its use in the repair of mechanical parts with high initial cost. These processes improve on the existing techniques of repair by material addition through welding and subsequent machining. Thus, the AM techniques allow a more precise control of the repair process, increasing the quality of the final part in terms of surface finish, dimensional accuracy and structural integrity [14]. Given the greater rigidity of use of the other additive manufacturing technologies, direct energy deposition (DED) processes stand out in their application in maintenance tasks.

17.4.7 Resource for Teaching and Learning

3D printing is a very powerful tool in teaching and learning as resource for teaching and learning theoretical concepts that are difficult to visualize, and specially in distance learning education of technical subjects. Some examples are the design of machines, equipment, and tools in manufacturing processes, such as stamping or extrusion dies, or the Bravais crystal lattices in Materials Science and Technology [15]. This is confirmed by the increase in teaching innovation projects where additive manufacturing plays a fundamental role.

17.4.8 Design and Manufacturing of Prototypes

Product and component prototypes are necessary in order to validate designs from different points of view. The marketing department, for example, uses prototypes to test different aspects of the products to the consumer before launching them on the market. On the other hand, designers also use prototypes to validate designs for new products or improvements to them. For this purpose, the use of AM for rapid prototyping is already stablished in most sectors, since it is not necessary to make a large investment to produce a few prototypes and it is also possible to quickly manufacture multiple design variants to evaluate the design [16, 17].

17.4.9 Potential Against Depopulation

Additive technologies have the potential not only to revolutionize the manufacturing of products, but also to create new environments for innovation and promote the entrepreneurial spirit of entrepreneurs in areas far from the currently established industrial poles. Thus, according to this point of view, the impact of AM on depopulation could be measured around some aspects, where AM can have an active role: (a) Relocation of production, (b) Access to in-situ diagnostic tests and (c) Creation of new job positions. In this line, there are several initiatives, among which it is worth highlighting at the national level (in Spain) the project [18].

17.5 Conclusions

This paper presents a set of dimensions with potential application of additive manufacturing, which allows us to understand the need for comprehensive training in this field and its role in the era of digitization and education in STEAM subjects (science, technology, engineering, art and mathematics). To this aim, both general and specialized training are necessary according to the profile and needs, as described in this paper.

Acknowledgements This work has been financed thanks to the UNED Teaching Innovation Project "Multi-criteria analysis for the improvement and expansion of the teaching of additive manufacturing contents in technical education with the blended learning methodology characteristic of

UNED" and has been realized in the frame of the projects and "Metodología multidisciplinar basada en proyectos para el aprendizaje activo en torno a las oportunidades que ofrece la fabricación aditiva" and "Evaluación del impacto de la implementación de herramientas virtuales y aplicaciones interactivas en remoto sobre la docencia a distancia en Ciencia y Tecnología de Materiales" of the GID2016-28 Teaching Innovation Group; and the Project "Additive Manufacturing as a technological disruption to fight against rural depopulation and social and spatial inequalities" (Project PLEC2021-007750 financed by MCIN/AEI /10.13039/501100011033 and by the European Union NextGenerationEU/ PRTR).

References

1. Rodríguez-Panes, A., Claver, J., Camacho, A.M., Sebastián, M.A.: Análisis normativo y evaluación geométrica de probetas para la caracterización mecánica de piezas obtenidas por fabricación aditiva mediante FDM. Anales de Ingeniería Mecánica 546–556 (2018)
2. Equipo de Investigación en Fabricación Aditiva – UNED. https://blogs.uned.es/fabricacionaditiva-additivemanufacturing/. Accessed 04 Nov 2022
3. Real Decreto 280/2021, de 20 de abril, por el que se establece el Curso de especialización en Fabricación aditiva y se fijan los aspectos básicos del currículo (2021)
4. International Organization for Standardization: ISO 17296-2:2015, Additive Manufacturing General Principles. Part 2: Overview of Process Categories and Raw Materials, Geneva (2015)
5. Programa Modular/Máster en Fabricación Aditiva. https://formacionpermanente.uned.es/tp_actividad/actividad/fabricacion-aditiva
6. Plataforma CSIC para el desarrollo de la Fabricación Aditiva FAB3D. https://pti-fab3d.csic.es/. Accessed 04 Nov 2022
7. Vaneker, T., Bernard, A., Moroni, G., et al.: Design for additive manufacturing: framework and methodology. CIRP Ann. **69**, 578–599 (2020)
8. Wiberg, A., Persson, J., Ölvander, J.: Design for additive manufacturing – a review of available design methods and software. Rapid Prototyp. J. **25**, 1080–1094 (2019)
9. Sotomayor, N.A.S., Caiazzo, F., Alfieri, V.: Enhancing design for additive manufacturing workflow: optimization, design and simulation tools. Appl. Sci. **11** (2021)
10. Pradel, P., Zhu, Z., Bibb, R., Moultrie, J.: A framework for mapping design for additive manufacturing knowledge for industrial and product design. J. Eng. Des. **29**, 291–326 (2018)
11. García-Dominguez, A., Claver, J., Sebastián, M.A.: Optimization methodology for additive manufacturing of customized parts by fused deposition modeling (FDM). Application to a shoe heel. Polymers (Basel) **12** (2020)
12. Yang, L., Harrysson, O.L.A., Cormier, D., et al.: Design for additively manufactured lightweight structure: a perspective. In: Proceedings of 27th Annual Internationa; Solid Free Fabr Symposium - An Additive Manufacturing Conference, pp. 2165–2180 (2016)
13. Rodríguez-Prieto, A.: Ingeniería inversa y caracterización avanzada de materiales para el establecimiento de requisitos de aceptación en procesos singulares de dedicación. Nucl España **402**, 55–58 (2019)
14. Bennett, J., Garcia, D., Kendrick, M., et al.: Repairing automotive dies with directed energy deposition: industrial application and life cycle analysis. J. Manuf. Sci. Eng. Trans. ASME **141** (2019)
15. Ayllón, J., Rodríguez-Prieto, A., García-Domínguez, A., et al.: Aplicación de la Fabricación Aditiva como herramienta de apoyo a la docencia en ciencia y tecnología de materiales. In: Cuenca V.P., Sáiz B. (eds.) Libro de actas del Congreso Universitario de Innovación Educativa en Enseñanzas Técnicas (CUIEET_29), pp. 179–213. UPV Valencia (2022)
16. Chen, S.E., Shanmugam, N.S., Boyajian, D., et al.: Prototyping rail track for micro-people movers using additive manufacturing: failure topology characterization. Constr. Build Mater. **281**, 122623 (2021)

17. Robinson, D.K.R., Lagnau, A., Boon, W.P.C.: Innovation pathways in additive manufacturing: methods for tracing emerging and branching paths from rapid prototyping to alternative applications. Technol. Forecast. Soc. Change **146**, 733–750 (2019)
18. Disrupting depopulation: Additive Manufacturing as technological disruption to fight against rural depopulation and social and spatial inequalities (Proyecto PLEC2021-007750 financiado por MCIN/AEI/10.13039/501100011033 y por la Unión Europea NextGenerat

Part III
Mechanism and Machine Science in the Engineering Program: Virtual Labs

Chapter 18
Matlab App for Teaching Planar Mechanism Kinematics

Alejandro Bustos, Higinio Rubio, and Juan Carlos Garcia-Prada

Abstract This paper explains a computational tool for the kinematic analysis of a complex mechanical system. This software, implemented in MATLAB, was designed to complement the teaching of undergraduate students in mechanical engineering in the field of mechanism design. The basic mechanism is an articulated system of links used in the leg of the biped robot Mimbot. The software consists of implementing a series of parametric models of the leg of the biped and allows detailed knowledge of the kinematic behaviour of the leg according to each of the model parameters to be analyzed. The program contains 5 modules, with different levels of difficulty, from the basic mechanism (Tchebyshev's mechanism coupled to a pantograph mechanism and 1 degree of freedom) to a complex mechanism with 3 degrees of freedom and two linear actuators. With this software, the students will learn to model the equations defining motion trajectory, and identify their corresponding graphs. In addition, this software aims to show students a compact, robust, friendly way, exploring the possibilities that certain variations or adjustments in some kinematic parameters of the mechanism can provide in the setting of other kinematic variables.

Keywords Matlab app · Mechanism kinematics · Raven's method · Practical learning

18.1 Introduction

A wise Chinese sage once said: "I hear and I forget. I see and I remember. I do and I understand". The concepts of Mechanism and Machine Theory courses are not always easily understood by the students, and implementing the practical learning hidden behind that proverb could improve their success. In particular, computer applications

A. Bustos (✉) · J. C. Garcia-Prada
Universidad Nacional de Educación a Distancia, c/ Juan del Rosal 12, 28040 Madrid, Spain
e-mail: albustos@ind.uned.es

H. Rubio
Universidad Carlos III de Madrid, Av. De La Universidad 30, 28911 Leganes, Spain

J. C. García Prada et al. (eds.), *Trends in Educational Activity in the Field of Mechanism and Machine Theory (2018–2022)*, Mechanisms and Machine Science 128,
https://doi.org/10.1007/978-3-031-25730-8_18

make the students' learning easier and give them a more comprehensive view of the problem. In addition, these applications allow personalizing difficult problems if we deal with them in a general view, achieving greater teaching effectiveness.

There are several examples in the scientific literature of interactive tools developed for the teaching of Mechanical courses [1–6]. In some cases, the interactive tool is just a set of tests embedded in an Android app, but other developments are true computer software. WinMec [2] is a useful tool for creating and analysing planar mechanisms by just adding and setting up the mechanism elements. The software tool developed in [1] allows the creation of 3D mechanisms and the computation of the kinematics and workspaces, among other features.

MATLAB is a powerful tool and programming language that every Engineering student should know. Having this in mind, it makes sense to develop educational-oriented apps based on MATLAB. This idea has been carried out in different engineering fields [3, 7–9] thanks to the tools provided by MATLAB to create attractive applications using interfaces and the computation capabilities of this programming language.

In this work, we present an app developed in MATLAB using GUIDE for the computation of the kinematics of a biped robot that is intended to bring practical learning to the classrooms of Mechanism and Machine Theory.

18.2 Raven's Method

Raven's method [10] is a useful and easy-to-understand tool for solving four-bar linkage planar mechanisms. It is based on the creation of a closed vector loop equation in which the vectors are replaced by complex numbers. By solving this equation, the position field is obtained. If the equation is differentiated two times, the velocity and acceleration fields are obtained.

To solve the equation, the vectors are replaced by complex numbers. After separating the real and the imaginary parts, a linear equation system composed of two scalar equations is obtained. By solving this system, the unknowns are obtained. Obviously, as only two equations are obtained, only two unknowns can be solved. Usually, the unknowns are the angles of the coupler and output links. This is the main limitation of the method and implies that solving mechanisms with more than four linkages requires the definition of more than one vector loop.

Figure 18.1 shows a typical four-bar linkage mechanism in which the linkages have been replaced by vectors to create the closed loop, which leads to Eq. (18.1). Replacing the vectors with their complex numbers leads to Eq. (18.2), in which the dimensions r_1, r_2, r_3 and r_4 are known, as well as angles θ_1 and θ_2. The unknowns are the angles θ_3 and θ_4.

$$\overrightarrow{r_2} + \overrightarrow{r_3} - \overrightarrow{r_4} - \overrightarrow{r_1} = \overrightarrow{0} \tag{18.1}$$

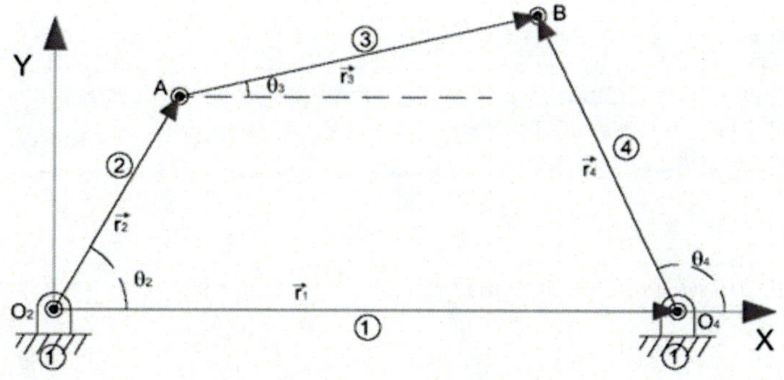

Fig. 18.1 Vector closed loop in a four-bar linkage mechanism

$$r_2 \cdot e^{j\theta_2} + r_3 \cdot e^{j\theta_3} - r_4 \cdot e^{j\theta_4} - r_1 \cdot e^{j\theta_1} = 0 \tag{18.2}$$

18.3 Mechanism Description

Mimbot robot is an evolution of the biped robot Pasibot [11, 12] (see Fig. 18.2), consisting of a mechanical system with one degree of freedom that can emulate the human gait. Biped Pasibot is based on a combination and adaptation of classical mechanisms: Tchebyshev's mechanism, pantograph mechanism and stabilization systems.

In addition to some dimensional adjustments, improvements added to biped Mimbot, regarding biped robot Pasibot, are two:

- A new stabilization system to ensure the foot will be always parallel to the ground, without additional actuation.

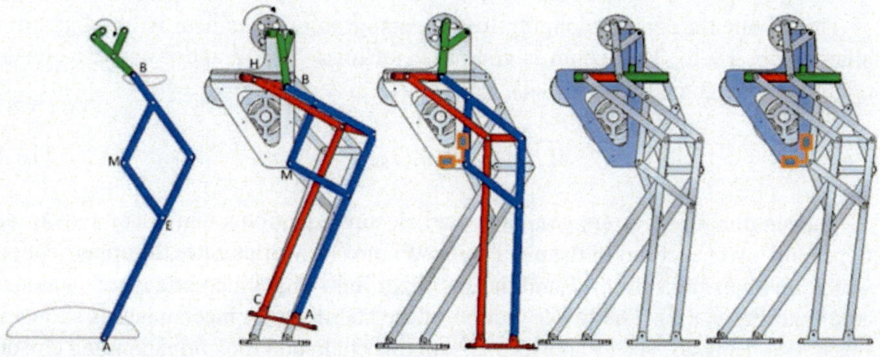

Fig. 18.2 Five steps in the evolution of MIMBOT

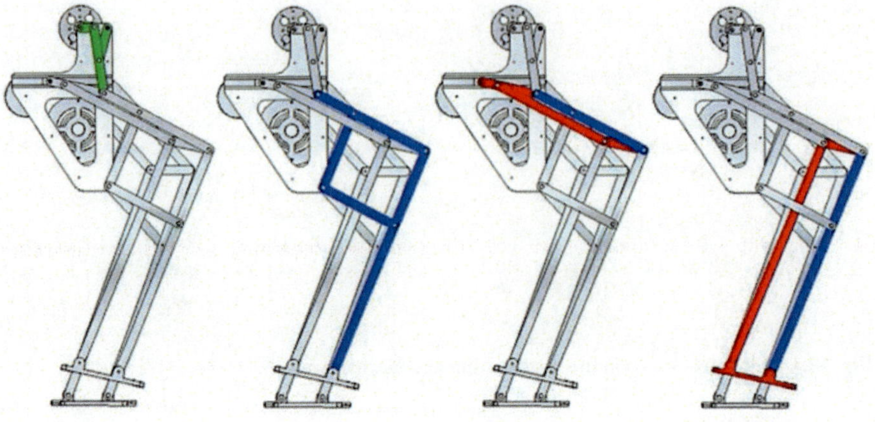

Fig. 18.3 Kinematic chains used in the parametric model. **a** Tchebyshev mechanism, **b** pantograph mechanism, **c** upper stabilization system, **d** lower stabilization system

- The addition of two linear actuators on each leg gives the robot the ability to develop similar skills to human ones. This way, it is possible to model the foot trajectory as desired, in order to lengthen or shorten the biped's gait, go up or down a step, or overcome a small obstacle. It also increases the degrees of freedom up to five (two more degrees of freedom in each leg).

In order to solve the kinematics of the biped robot, four kinematic chains are defined according to Fig. 18.3. The first kinematic chain (in green in Fig. 18.3a) is associated with Tchebyshev's mechanism and includes links 1, 2, 3 and 4. The closure vector equation for this chain is given by Eq. (18.3)

$$r_1 e^{j\theta_1} + r_2 e^{j\theta_2} + r_3 e^{j\theta 3} + r_4 e^{j\theta_4} = 0 \tag{18.3}$$

The kinematic chain belonging to the pantograph mechanism is highlighted in blue in Fig. 18.3b. This chain is composed of links 5, 6, 7 and 8 and the closure relationship is given by Eq. (18.4).

$$\overrightarrow{MB} + r_6 e^{j\theta_6} + r_8 e^{j\theta_8} = 0 \tag{18.4}$$

The stabilization system generates two closure equations, which come from the upper and lower sections of the mechanism. Figure 18.3c illustrates the upper section, which involves links 5, 9, 11 and the stabilizer link. This kinematic chain is associated with Eq. (18.5). The lower section of the stabilization mechanism is shown in Fig. 18.2d. Links 6, 10, 11 and 12 make up this chain and the corresponding closure equation is defined by Eq. (18.6).

$$r_C e^{j\theta 7} + r_{11} e^{j\theta 11} = r_f e^{j\theta 8} + r_9 e^{j\theta 9} \tag{18.5}$$

$$r_A e^{j\theta 8} + r_{12} e^{j\theta 12} = r_{11} e^{j\theta 11} + r_{10} e^{j\theta 10} \tag{18.6}$$

Knowing the angular position of linkage 2, which is attached to the motor and by solving Eqs. (18.3)–(18.6), the angles of every linkage can be obtained. As the lengths of the linkages are known, the position of every point of the mechanism can be deduced easily. The whole kinematics of the mechanism is obtained by differentiating Eqs. (18.3)–(18.6) two times.

18.4 App Description

The fact of using classical mechanisms in the design of the Mimbot mechanical system allows using this mechanical system as a real example and basis for teaching vectorial mechanics. To fulfil this role, Mimbot 5.0 app is developed. It makes easier the teaching of exercises and allows verifying that the result of solving the posed theoretically in lessons matches with those provided by the tool.

In addition, if the real-scale biped prototype dimensions are used, results from the software tool (similar to those obtained from the hand solution of equations) can be compared with those measured by sensors on the prototypes. This utility makes the computer application Mimbot 5.0 a powerful tool also for research, as it makes the dimensional characterization of robot design easier for its functional optimization.

The app is developed using GUIDE (Graphical User Interface Development Environment), a MATLAB tool specially developed to create GUIs. Using this tool, the design and presentation of GUIs can be easily and quickly built: just selecting, pulling, dragging and customizing the properties of buttons, sliders, textboxes, etc. [13]. Also, the fact of being part of MATLAB gives us the chance to take advantage of the familiarity that engineering students have with MATLAB.

18.4.1 Structure and Implementation

The app is structured in a welcome window and three operating windows. The first operating window allows the user to select the step in the MIMBOT evolution to analyse. The second window allows the introduction of data for computing the robot's kinematics, and the third and last window shows the results of the computations. The detailed flowchart of the app is shown in Fig. 18.4.

The block "Module selection" opens the window shown in Fig. 18.5 that allows selecting one of the five available modules for computing the kinematics of the mechanism. Each module corresponds to one step in the evolution of the biped robot which is illustrated in Fig. 18.2.

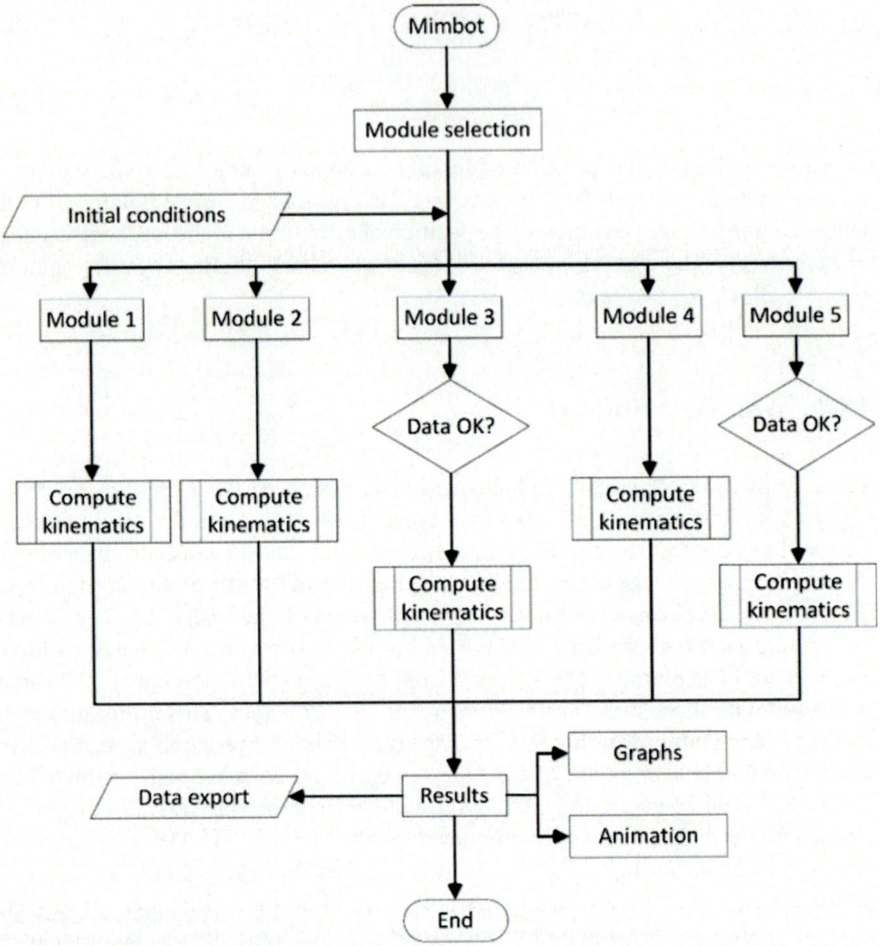

Fig. 18.4 Flowchart of the app

After that, the user is asked to introduce all the input data required for the computation of the robot's kinematics. A new window opens for setting the lengths of the linkages, the temporal parameters and the motion of the linear actuators if needed. Five different windows are created to cover the five modules (evolution steps) available, although all are similar. The left area of the window is reserved for dimensional and temporal data. The centre/right area shows a drawing of the mechanism numbering all the linkages and the dimensions. The far right area is only available in modules 3 and 5 and is reserved for inputting the parameters of the linear actuators.

Figure 18.6 shows the window for the third module, which is the original design of the robot with the addition of the linear actuators. By pressing the "Doubled" button it is possible to introduce the dimensional data for the stabilization mechanism. By pressing again this button the window returns to the input of the Tchebychev and

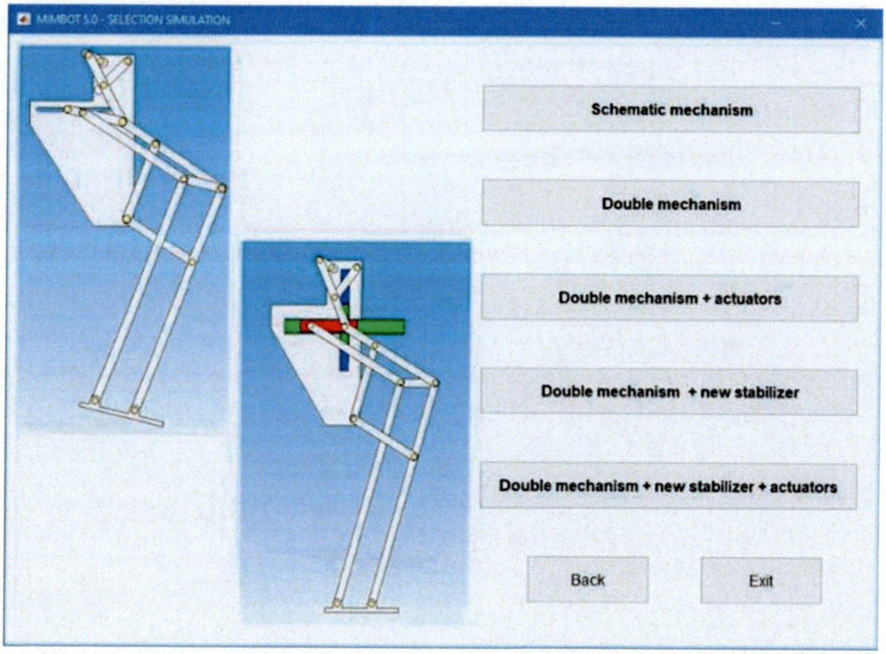

Fig. 18.5 Module selection window

pantograph mechanisms. In modules 3 and 5, the student can also check the motion of the linear actuators by clicking on the "Actuator graphics" menu.

Once all input data are introduced and checked, the student must press the "Calculate" button to compute the kinematics and the button "Next" will highlight once this computation finishes.

There are several options for setting the motion of the linear actuators:

- Fixed or null career (both actuators): A fixed or null displacement of point M is set up. The actuator remains inactive during the simulation. This feature allows modifying the global centre of mass within the support area defined by the feet.
- Front gait modification (horizontal actuator): It modifies the path of the foot to adjust the length of the step at its front, according to a predetermined path feature designed for this purpose. If this option is selected, a slider bar to select the length of the actuator, within a limited range, will appear. Depending on the value entered an elongation step (for positive values) or shortening (for negative values) will be got. If the value entered is zero, the horizontal actuator is inactive.
- Rear gait modification (horizontal actuator): same motion as the previous one, but in this case, the variation is at the back of the step.
- Foot elevation (vertical actuator): it consists in modifying the maximum height of the foot during the flight phase. The only input parameter needed is the actuation distance. A positive value will raise the foot and a negative one will lower it.

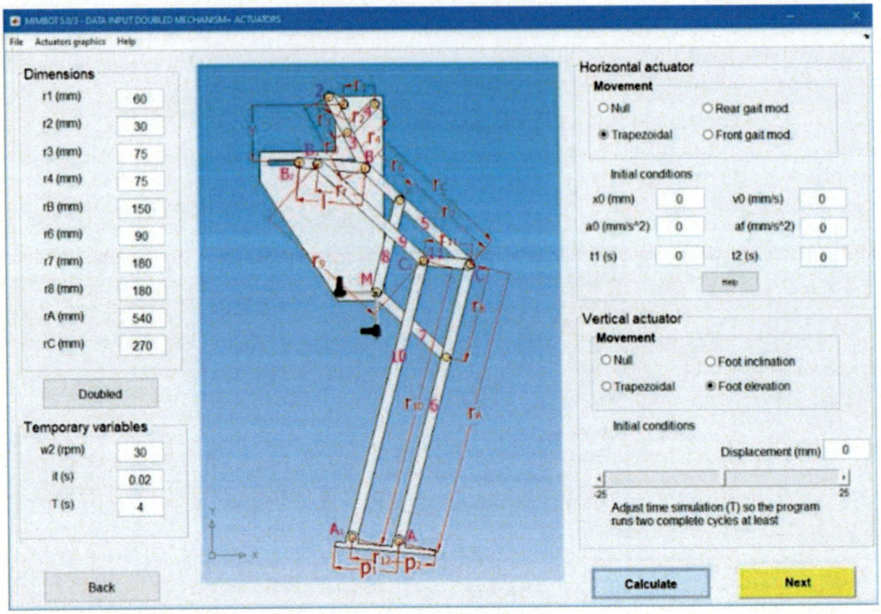

Fig. 18.6 Data input and computation window

- Foot inclination (vertical actuator): it is also a change in the height of the foot, but this time it takes place at the final phase of flight. In this movement, the foot is elevated during the whole phase of flight and returns to the original cycle during the support phase. Positive values will elevate the foot and negative values will fall the foot.
- Trapezoidal (both actuators) The last predefined movement is a combination of two uniformly accelerated linear motions (at the beginning and end of the actuator stroke) and a uniform linear motion in the middle section. Therefore, it requires setting up the initial position, velocity and acceleration as well as the acceleration for the last section and the time instants in which the change from one motion to another takes place.

The results window is customized for every module too as the number of results varies for each mechanism. The aspect is similar in all cases and the available results are grouped by mechanism type and position, velocity or acceleration. In order to visualize the results, the student must select the graph they want to analyse from the list and click on "Draw graphic". "Links" and "Angles" buttons open auxiliary windows to remember the naming of links and angles. The "Animation" button opens a new window that allows visualizing the motion of the mechanism (Fig. 18.7).

Each animation window has 7 buttons, that perform the typical operations reproducing a video. The fast and slow rewind and fast and slow forward buttons work only when the animation is running. If the animation is stopped, either because the

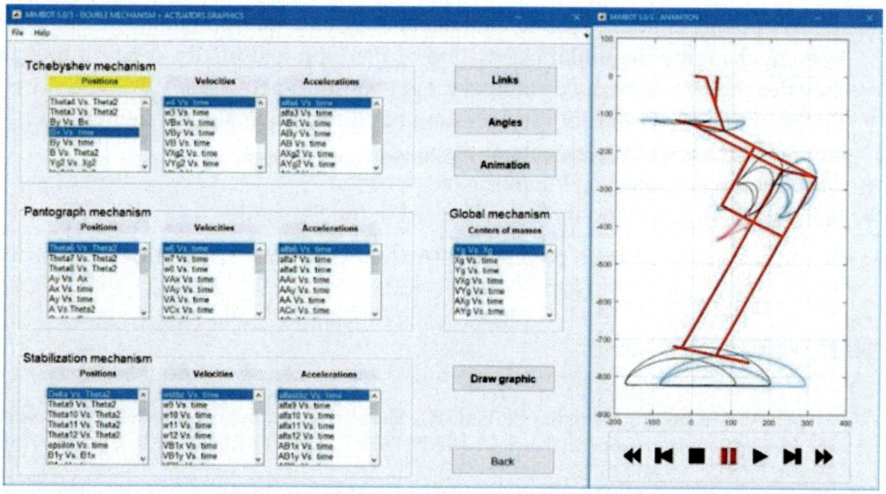

Fig. 18.7 Results and animation windows

stop or pause buttons are pressed or because it is finished, these buttons have no effect, being necessary to press the play button first.

The animation window also shows the trajectory of the most characteristic points of the mechanism (B, B1, B2, C, C1, A, and A1) as well as the trajectory of the whole centre of mass. Trajectories corresponding to points B, B2, C and A (Techebyshev and pantograph mechanisms) are drawn in blue, while those corresponding to points B1, C1 and A1 are painted in black, and the whole centre of mass is painted in pink.

18.5 Conclusions

A MATLAB app is developed using GUIDE for the computation of the kinematics of a biped robot that is composed of three basic mechanisms: a Tchebyshev‘s mechanism, a pantograph mechanism and a stabilization mechanism. The equations to solve the whole mechanism are posed based on Raven's method, which is taught in undergraduate courses at Universidad Carlos III de Madrid.

The educational nature of the tool lies in its visual design and ease of use, which allows students to try several robot configurations based on three basic mechanisms and to know the behaviour that such configurations can generate.

The software is characterized by:

- Providing a friendly and intuitive graphic interface very easy to use.
- Allowing 5 study levels, as it contains 5 modules with different difficulty levels: the first one consists of a Tchebyshev's mechanism coupled to a pantograph mechanism and 1 DOF; the second one includes the stabilization system; the third one considers the inclusion of two actuators adding 2 DOF to the system;

the fourth one, similar to the second one, develops a variation in the stabilization system, changing the initial orientation of the foot; and finally, the fifth module includes the new horizontal stabilizer and two linear actuators.

- Representing the graphs of the evolution of all the kinematic parameters and the centres of mass of the elements of the model.
- Ability to save, export and import data.
- Animation of geometric models proposed in the 5 modules.

References

1. Petuya, V., Macho, E., Altuzarra, O., Pinto, C., Hernandez, A.: Educational software tools for the kinematic analysis of mechanisms: KINEMATIC SOFTWARE TOOLS. Comput. Appl. Eng. Educ. **22**, 72–86 (2014). https://doi.org/10.1002/cae.20532
2. Bataller, A., Ortiz, A., Cabrera, J.A., Nadal, F.: WinMecC: software for the analysis and synthesis of planar mechanisms. In: Wenger, P., Flores, P. (eds.) New Trends in Mechanism and Machine Science, pp. 233–242. Springer International Publishing, Cham (2017)
3. Pena, P., Utschig, T., Tekes, A.: Reinforcing student learning by MATLAB simscape GUI program for introductory level mechanical vibrations and control theory courses. Int. J. Mech. Eng. Educ. **50**, 849–868 (2022). https://doi.org/10.1177/03064190221085038
4. González, D., Estrada, E., Roldán, J.: Android application for the study of four-link planar mechanisms. Entre Ciencia e Ingeniería. **10**, 41–51 (2016)
5. Corral Abad, E., Gómez García, M.J., Ruiz Blázquez, R., Bustos-Caballero, A., García-Prada, J.C.: Using an android app for teaching mechanism and machine theory. In: García-Prada, J.C., Castejón, C. (eds.) New Trends in Educational Activity in the Field of Mechanism and Machine Theory, pp. 103–110. Springer International Publishing, Cham (2019)
6. Urbina Coronado, P.D., Demeneghi, J.A.A., Ahuett-Garza, H., Orta Castañon, P., Martínez, M.M.: Representation of machines and mechanisms in augmented reality for educative use. Int. J. Interact. Des. Manuf. **16**, 643–656 (2022). https://doi.org/10.1007/s12008-022-00852-x
7. Sardiñas-Fernández, R., García-Juárez, A., Zaldívar-Huerta, I.E., Abril-García, J.H.: MATLAB app designer tool to study a microwave photonic filter that integrates analog and digital modulation formats. Optik **202**, 163589 (2020). https://doi.org/10.1016/j.ijleo.2019.163589
8. Wong, S.Y., Lim, S.Y.: Electromagnetic education: development of an interactive GUI for demonstrating wave polarization. Comput. Appl. Eng. Educ. **28**, 1190–1219 (2020). https://doi.org/10.1002/cae.22296
9. Battestilli, L., Silverberg, L., Eischen, J., Thomas, C.: CADApps: MATLAB Apps for Core Courses in Aerospace and Mechanical Engineering Curricula, vol. 8 (2019)
10. Simón Mata, A., Bataller Torras, A., Cabrera Carrillo, J.A., Ezquerro Juanco, F., Guerra Fernández, A.J., Nadal Martínez, F., Ortiz Fernández, A.: Analytical methods for the kinematic analysis of planar linkages. Raven's method. In: Fundamentals of Machine Theory and Mechanisms, pp. 111–141. Springer International Publishing, Cham (2016)
11. Rubio, H., Meneses, J., Castejón, C., Jardón, A., Giménez, A., García-Prada, J.C.: Mechanical design of walking robot pasibot-parametric model and gait analysis. In: Mobile Robotics, pp. 701–709. WORLD SCIENTIFIC, Istanbul, Turkey (2009)
12. Meneses, J., Castejón, C., Corral, E., Rubio, H., García-Prada, J.C.: Kinematics and dynamics of the quasi-passive biped "PASIBOT." SV-JME **57**, 879–887 (2011). https://doi.org/10.5545/sv-jme.2010.210
13. Smith, S.T.: MATLAB: advanced gui development. Dog Ear, Indianapolis (IN) (2006)

Chapter 19
General-Purpose Software Tools in Teaching MMS

Jose Luis Torres-Moreno, Jose Luis Blanco-Claraco, and Antonio Gimenez-Fernandez

Abstract Nowadays, teaching of MMS involves the use of multimedia and software tools and that facilitate the understanding of theoretical contents and the development of practical activities. In recent years, specific purpose tools have been developed for this purpose. However, since the constant improvement of general-purpose programs in the field of mathematics or computer-aided design, which increasingly include more toolboxes and more intuitive interfaces, this paper presents novel training methodologies based on the use of these types of programs. Specifically, the use of GeoGebra, MATLAB/Simulink and Autodesk Inventor is proposed. In view of the results obtained after the first test courses, the teaching experience can be considered satisfactory and with great potential for the future.

Keywords Teaching · Software · Graphical-methods · CAD

19.1 Introduction

The subjects related to Mechanism and Machines Science (MMS) constitute a fundamental basis in any Mechanical Engineering degree. Since its origins during the nineteenth century, this discipline has been characterized by the combination of mathematics and geometry [1]. Given that its foundations are based on the study of motion, the entire bibliography and methodology is supported by numerous diagrams, schemes, or graphics that try to illustrate the geometric and physical laws governing such motion. Despite the advances in analytical and computational calculation methods that occurred at the end of the twentieth century, graphical methods are still considered a highly didactic approach for learning MMS. In addition, they also benefit from the advancement of computing since, as will be seen in the next section, the development of interactive geometry tools.

J. L. Torres-Moreno (✉) · J. L. Blanco-Claraco · A. Gimenez-Fernandez
University of Almeria, Carretera de Sacramento S/N, 04120 Almeria, Spain
e-mail: jltmoreno@ual.es

J. C. García Prada et al. (eds.), *Trends in Educational Activity in the Field of Mechanism and Machine Theory (2018–2022)*, Mechanisms and Machine Science 128,
https://doi.org/10.1007/978-3-031-25730-8_19

MMS is generally a discipline that involves advanced spatial vision skills and is a significant challenge for students, especially when they are faced with this discipline for the first time. For this reason, since the appearance of computers, the use of software tools has been promoted, including animations and interactive apps that facilitate the assimilation of content and the resolution of problems. Some of these programs have been developed on the premise that general purpose programs are more expensive and difficult to use for teaching tasks. In this sense, there are programs such as WinMecC [2], GIM [3], SAM [4], MechAnalyzer [5, 6] or even approaches based on augmented reality, such as ARMES [7].

However, with the improvement of general-purpose programs in the field of mathematics or computer-aided design (CAD), and the ease of digital learning of the new generations of students, more familiar with the use of software, this kind of programs becomes suitable for teaching MMS. This is the approach of the present work, where not only training strategies based on open source programs are proposed, as was done with GeoGebra in [8], but another series of strategies and activities are proposed with commercial software as MATLAB/Simulink and Autodesk Inventor. In view of the preliminary results obtained, it can be considered that the teaching experience has been satisfactory, both for learning MMS and with the integration of the knowledge acquired in other subjects of the degree. The rest of the article is structured as follows: Sect. 19.2 presents the selected tools and the contents where their use best fits. In Sect. 19.3 some proposed activities are presented where students use the tools presented in Sect. 19.2. Finally, in Sect. 19.4 the conclusions are presented, and future works are proposed.

19.2 The Selected Tools and Their Applications

This section presents the use of three software tools, one of them open source and the other two commercial. The advantages of using open-source tools are obvious, especially if they belong to active projects, as is the case with the chosen option. However, access to the other two tools that are proposed does not represent a significant entry barrier for their use in teaching in MMS, since they are general purpose programs that usually include agreements with numerous education centers, since they are transversal and widely used in many other subjects. In addition, they include programs with licenses for students, which facilitates their access. The first of the tools considered focuses on the graphic method for teaching kinematics and the synthesis of mechanisms, as well as an introduction to dynamics, all using graphic methods. It is the open-source program GeoGebra, which is available for computers and mobile devices. The second is MATLAB/Simulink, which can help to understand the kinematics and dynamics of mechanisms by simplifying the tedious algebraic calculations involved. Finally, Autodesk Inventor is presented as an interesting option for dynamic simulation and for the design and calculation of gears and cams.

19.2.1 GeoGebra

GeoGebra is an open-source software that allows, among other functions, to perform algebraic calculations and represent mathematical functions in 2D, in an interactive way, and with a simple and intuitive interface. In addition, it incorporates a series of functions that allows to work with vectors, circumferences, segments, or straight lines, making it possible to impose conditions of perpendicularity and parallelism. On the other hand, it allows taking measurements of angles and distances and storing all these parameters in variables that can be used to carry out subsequent calculations, in a completely parameterized way. This is an essential aspect when implementing the graphical method for calculating velocities and accelerations based on diagrams.

GeoGebra has been used satisfactorily in teaching sessions of the Mechanisms Theory subject, of the third semester of the Degree in Mechanical Engineering, at the University of Almería. Below are some contents for which it has been used.

Introduction to the Basic Types of Mechanisms

Since it is the first subject of the degree in which the students face the modeling of mechanisms, they are not familiar with the symbolic notation of Franz Reuleaux [9]. Therefore, it is especially useful to count, in the first introductory sessions, with animations of the articulated quadrilateral mechanism, the slider-crank mechanism, the inverted slider-crank mechanism, the quick-return mechanism, and different types of 6-bar mechanisms.

19.2.1.1 Proof of the Grashof's Theorem

From the determination of the lengths of the bars that compose an articulated quadrilateral, it is visually checked whether it is a crank-rocker, double crank, or double rocker mechanism.

19.2.1.2 Instantaneous Centers of Rotation (ICR)

Thanks to the facility to draw lines parallel and perpendicular to the bars of a mechanism designed in GeoGebra, the location of its ICRs is direct, even if its location is far from the links. In addition, the velocities of any point of this can be calculated from the data corresponding to the degrees of freedom, since calculations can be implemented that use the results of the distances of the ICRs to the points considered as variables.

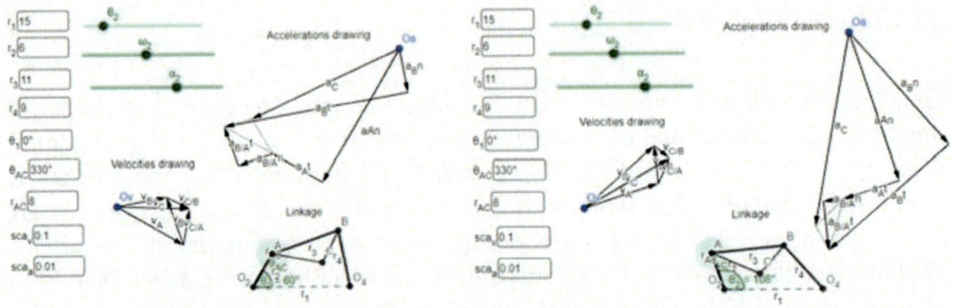

Fig. 19.1 GeoGebra template for the four-bar mechanism

19.2.1.3 Velocities and Accelerations Computation by Graphical Methods

An important part of the subject consists of understanding the graphical method of calculating velocities and accelerations. This method can be implemented in a completely parameterized way involving the geometric parameters of the mechanism, so that it is valid for any combination of these, in a given type of mechanism. As an example, three templates have been developed for the articulated quadrilateral, slider-crank and inverted slider-crank mechanisms. These templates can be used, for example, to check the results of exercises solved by hand or by other methods. Hence, these templates constitute a powerful tool for self-study. Figure 19.1 shows two time lapses of an animation of the template corresponding to the articulated quadrilateral. As can be seen, at each new position the velocity and acceleration kinematic diagrams are recalculated, considering the new directions and magnitudes of the velocity and acceleration vectors. The figure also shows the editable fields corresponding to the geometric parameters of the mechanism and the values of the degree of freedom.

19.2.1.4 Dynamics of Mechanisms by Graphical Methods

The use of GeoGebra facilitates teaching mechanism dynamics exercises since it allows to easily solve the static and dynamic equilibrium equations of each element that composes a mechanism graphically. This approach is also valid for the application the superposition principle to calculate the complete mechanism under the action of external forces. In each case, the forces polygon that leads to the resolution of the equations can be drawn.

19.2.1.5 Balancing of Rotors by Graphical Methods

In this case, a similar approach to the previous one is used in which the dynamic equations of an unbalanced rotor can be solved to calculate the mass and position of a balancing load. As in the previous case, the implementation of graphical methods can be carried out in an easy and simple way through GeoGebra.

19.2.1.6 Mechanisms Synthesis

Finally, the demonstration of the Freudenstein equations for function generation synthesis, the rigid-body guidance synthesis or the trajectory generation synthesis for three precision points may be addressed by using GeoGebra. In the first case, the software is used to check the result of a solution previously calculated by hand, while in the others the program is used to solve the problem using graphical methods.

19.2.2 MATLAB/Simulink

MATLAB/Simulink, from Mathworks software company, is widely known and used in numerous engineering applications. It has a large number of toolboxes, two of which are especially useful in teaching and research in MMS: The Symbolic Math Toolbox, and Simscape Multibody, which are suitable for the following task.

19.2.2.1 Kinematics of Mechanisms

In this context, the use of MATLAB/Simulink allows to calculate the problems of position, velocity and acceleration through the use of symbolic variables and algebraic operations with vectors in a simple and direct way. In addition, it allows to solve the kinematic problem of the Stephenson II mechanism, which is not possible by means of the standard version of GeoGebra since it contains two 5-bar loops.

19.2.2.2 Derivation of the Equations of Motion from the Lagrange's Method

As is well known, this method is based on the derivation of the Lagrangian function of a mechanical system. For this, it is necessary to carry out tedious partial derivatives with respect to generalized coordinates and velocities, as well as solving the resulting equations for the variables of interest and considering, where appropriate, the constraints. Thanks to the calculation power of the Symbolic Math Toolbox, these operations are carried out in a few seconds. In addition, using a MATLAB environment, allows the graphical representation of the results during a simulation.

19.2.2.3 Implementation of the Newton-Euler Method for the Dynamic Calculation of Mechanisms and Robots

In the case of planar mechanisms, for example, the dynamic calculation of a slider-crank mechanism involves solving a system of 8 equations with 8 unknowns. These equations can be easily generated from the calculation of the accelerations of the centers of gravity of the links that comprises the mechanism and their angular accelerations, following the approach presented in the previous section using the Symbolic Toolbox. Hence, their resolution is immediate, by means of techniques such as the matrix method [10]. For the 3D case, this method is characterized by a high recursive use of rotation matrices. The ease offered by MATLAB for the implementation of this type of operations makes it ideal for carrying out dynamic simulations of such systems.

19.2.2.4 Cams Design

The main cycloidal, harmonic, and polynomial cam profiles, as well as their derivatives, can be easily implemented in functions that take the cam angle as an input argument. Thanks to the powerful for graph representation, these diagrams can be displayed as functions of time or the cam angle, as well as the cam profile by means of the polar representation of the displacement diagram. As will be seen later, these results can be compared with those obtained experimentally with a model corresponding to a real cam-follower system, all on the same platform.

19.2.2.5 Dynamic Simulation of Mechanisms and Robots Using Simscape Multibody

Learning the rotational dynamics of the rigid body involves complex three-dimensional calculations and operations related to transformations of reference frames and rotation matrices. Thanks to the use of Simscape Multibody, animations involving these transformations can be easily carried out, which allows them to be viewed in a simple way, facilitating their understanding. In addition, through a correct definition of the inertial parameters and the joints between the elements of a simulation, dynamic analysis of robots and mechanisms can be performed. This analysis can be used to contrast the results obtained by the previously mentioned approaches based on the Lagrangian function and the Newton-Euler method.

19.2.3 Autodesk Inventor

Autodesk Inventor is a popular CAD/CAE software that also allows the implementation of CAM functions. Among its modules, it has one called Design Accelerator that allows to design and calculate different transmission systems such as cams or gears, as well as a module to perform Multibody Dynamics Systems (MBS) simulations. As is known, one of the most tedious tasks in performing a dynamic analysis is the definition of the inertial parameters of the elements that compose it. Except in the rare case that the analysis involves only elements with simple geometries, this step is usually carried out with CAD software. The fact that dealing with both modeling and dynamic analysis can be carried out under the same environment is a significant advantage over using other tools, considering the learning curve. Something similar occurs with the design and calculation of gears. Unlike other tools, Autodesk Inventor not only facilitates the design and simulation of gears, but also provides a detailed report on analysis of resistance, fatigue, efficiency, etc., which allows redesign, if necessary, the system in such a way that the transmission is optimized, all under the same platform. Finally, regarding the calculation of cams, the Design Accelerator module allows to design, calculate, and simulate cams in a simple and fast way. An interesting advantage offered by this program is the possibility not only to select standard profiles, but also to import custom profiles.

19.3 Proposed Activities

This section presents some proposals for activities in the subjects of Theory of Mechanisms and Machine Dynamics, of the third and fourth semesters of the Degree in Mechanical Engineering at the University of Almería, respectively. Both subjects comprise 6 ECTS and include 90 h of autonomous student work, including these activities.

19.3.1 Kinematic Analysis of a 6-Bar Mechanism

This activity consists of the modeling and calculation of a 6-bar mechanism belonging to the Stephenson or Watt kinematic chains, in which some revolution joints can be replaced by sliders. The work must be carried out in pairs of students. Each group will work on a different mechanism, which can be obtained from photographs or video frames of real or rendered mechanisms, as well as machinery diagrams or even patents. Thanks to the facility that GeoGebra offers to work on a background image (see Fig. 19.2), the modeling of the mechanism can be carried out even without having dimensions. With this activity two objectives are achieved: on the one hand, promotes the interest in the search for real applications of mechanisms is increased and tools are acquired to visualize the movement of real machines from a sufficiently representative image of the elements that define their kinematic chains. On the other hand, the knowledge about the procedures for calculating velocities and accelerations is consolidated, emphasizing graphical methods.

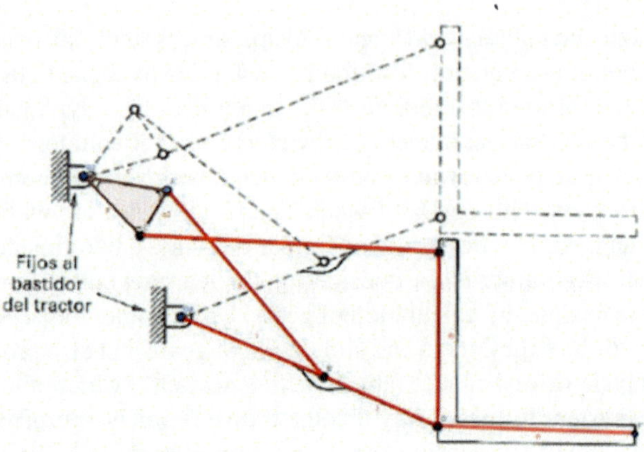

Fig. 19.2 Developing a 6-bar model from a picture

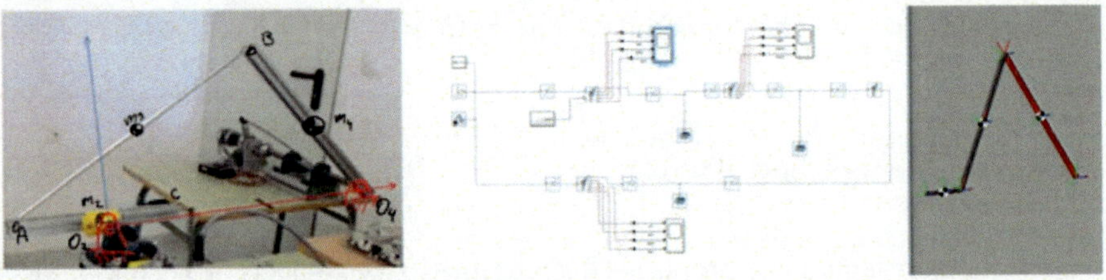

Fig. 19.3 Realt testbed and dynamic model

19.3.2 Dynamic Analysis of a Mechanism

In this activity, the modeling of a 4-bar mechanism based on the experimental testbed shown in Fig. 19.3 is proposed. For this task, the geometric and inertial parameters of its elements are provided, and two case studies are assumed: The first consists of a test by free fall in which the crank is decoupled from the DC motor, and in the second the motor is configured to provide the crank with constant speed motion. Students must implement the dynamic equations in a MATLAB script and perform two simulations: (i) a direct dynamic simulation in which the motion described by the mechanism is calculated from the gravitational forces, and (ii) an inverse dynamic simulation in which the torque provided by the DC motor to produce the registered movement is calculated. In both cases, the Lagrange method with constraints is used, and Simulink is used to carry out the simulation. This is especially useful for integrating the equations of motion in the first case, and for directly checking the results of both cases with those obtained using Simscape Multibody.

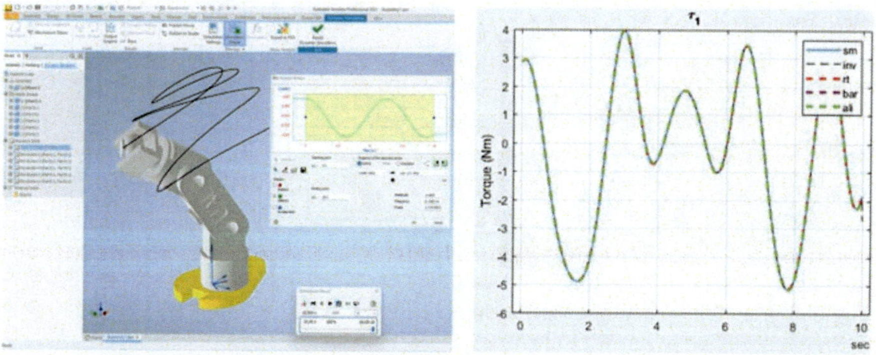

Fig. 19.4 Robot dynamic simulation

19.3.3 Dynamic Analysis of a Manipulator Robot

This activity consists of the modeling and dynamic analysis of a manipulator robot using MATLAB/Simulink and Simscape Multibody. As starting data, an Autodesk Inventor project is provided that contains both the complete CAD model of the robot and an inverse dynamic simulation from a trajectory generated by polynomial functions in each joint. Since the joints motion laws are known, these will constitute the input data of the models to be built by the students. Furthermore, all dimensional measurements and inertial parameters can be taken directly from Autodesk Inventor to be defined in the script to be implemented. In the case of the model in Simscape Multibody, the links can also be imported directly in .step format. In this way, the results obtained both with the implementation of the Newton-Euler method and with Simscape Multibody can be compared with those obtained with Autodesk Inventor. Figure 19.4 shows this comparison, where two variants of the method and the results obtained using the Robotic Toolbox have also been considered, as an additional element of validation. It should be noted that, for simplicity, only those of one of the 6 joints of the robot are shown, being the results equally precise in the rest.

19.3.4 Cam-Follower System Design

For the development of this activity, a partial cam displacement diagram is proposed for a given constant input velocity, and some boundary conditions of the cam-follower system. The idea is that a group of students (generally the group that completes the activity in the shortest time) can 3D-print the designed cam and it can be tested on the testbed shown in Fig. 19.5. The cam design is carried out using the Autodesk Inventor Design Accelerator, and its operation is validated with both Inventor and MATLAB/Simulink. Figure 19.5 shows also the results obtained by the theoretical model of the designed cam and the experimental data obtained on the testbed.

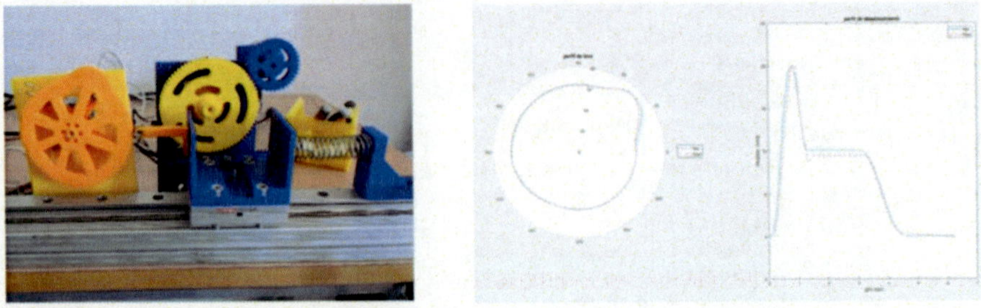

Fig. 19.5 Cam-follower system experiment

19.4 Conclusions

In this work the importance of the use of general-purpose software tools in the teaching of MMS has been highlighted, and three well-known programs have been proposed. To support its use, application cases have been exposed throughout the contents covered by two subjects taught in the Degree in Mechanical Engineering at the University of Almería, which range from the introduction to this discipline to the dynamic analysis of three-dimensional mechanisms. In addition, some of the activities that have been proposed to students in recent courses, which involve the use of these programs, have been presented. In view of the experience, the programs used play an important role both in the teaching of the contents and in the development of activities by the students, promoting motivation and interest in learning and making it possible to carry out more complex projects than if only worked with hand calculations. Another added advantage is that, as they are general-purpose tools, the learning curve is fast, and the knowledge acquired about these tools can produce benefits in other courses of the Degree. As future work, surveys are proposed to quantitatively evaluate the level of satisfaction with the use of these tools in subjects related to MMS.

Acknowledgements This work was partly funded by the projects UAL2020-TEP-A1991 (UAL/CTEICU/FEDER), PY20_00767 (CTEICU/FEDER), and the educational 20_21_2_07C of the University of Almeria.

References

1. Reuleaux, F., Kennedy, A.B.W.: The Kinematics of Machinery: Outlines of a Theory of Machines. Macmillan (1876)
2. Bataller, A., Ortiz, A., Cabrera, J.A., Nadal, F.: WinMecC: software for the analysis and synthesis of planar mechanisms. In: Mechanisms and Machine Science, vol. 43, pp. 233–242, Kluwer Academic Publishers (2017)
3. Macho, E., Urízar, M., Petuya, V., Hernández, A.: Improving skills in mechanism and machine science using gim software. Appl. Sci. (Switz.) **11**(17) (2021)

4. A. E. Software, SAM
5. Hampali, S., Chittawadigi, R.G., Saha, S.K.: "MechAnalyzer: 3D model based mechanism learning software. In:L 2015 IFToMM World Congress Proceedings, IFToMM 2015, pp. 25–30 (2015)
6. Chih-Ching, H., Chin-Hsing, K., Daisuke, M., Yukio, T.: Advances in Mechanism and Machine Science, vol. 73. Springer International Publishing (2019)
7. Wang, Y., Ong, S.K., Nee, A.Y.C.: Enhancing mechanisms education through interaction with augmented reality simulation. Comput. Appl. Eng. Educ. **26**(5), 1552–1564 (2018)
8. Iriarte, X., Aginaga, J., Ros, J.: Teaching mechanism and machine theory with GeoGebra. Mech. Mach. Sci. **19**(October), 211–219 (2014)
9. Moon, F.: Franz Reuleaux: contributions to 19th century kinematics and theory of machines. Appl. Mech. Rev. **56** (2003)
10. Mata, A.S., Bataller, A., Juan, T., Cabrera, A., Francisco, C., Juanco, E., Jesús, A., Fernández, G., Martínez, F.N., Ortiz Fernández, A.: Mechanisms and Machine Science 40 Fundamentals of Machine Theory and Mechanisms. (2016)

Chapter 20
A Distance Teaching Experience in Gear Design with Autodesk Inventor

Miryam B. Sánchez Sánchez, Miguel Pleguezuelos González, and José I. Pedrero Moya

Abstract Power transmission systems are subjected, under operating conditions, to diverse kinds of loads. Their failure is studied, from a theoretical point of view, in Machine Design subjects in Engineering Schools. This paper presents the procedure, proposed to students who study the subject through distance education, for modelling and analysing the stresses on a gear transmission, based on the Finite Element Method (FEM). Nowadays, FEM is a widely used tool for stress analysis and deflection in mechanical design. Considering the importance of this kind of tool, Autodesk Inventor is used to perform the analysis of a mechanical system. This analysis is a suitable complement to the explanations and theoretical developments presented during the teaching activity, serves as a support to student learning, and is used as part of the continuous evaluation.

Keywords Distance learning · Finite element method · Gear transmission

20.1 Introduction

Distance learning in higher studies is a challenge that requires the use of additional techniques and skills to those necessary to address similar studies in the face-to-face mode. The use of new communication technologies, adapted computer media and the application of more attractive and effective teaching methods, especially in the teaching of STEM subjects, represent a necessary aid for the achievement of teaching

M. B. Sánchez Sánchez · M. Pleguezuelos González (✉) · J. I. Pedrero Moya
Universidad Nacional de Educación a Distancia - UNED, Madrid, Spain
e-mail: mpleguezuelos@ind.uned.es

M. B. Sánchez Sánchez
e-mail: msanchez@ind.uned.es

J. I. Pedrero Moya
e-mail: jpedrero@ind.uned.es

J. C. García Prada et al. (eds.), *Trends in Educational Activity in the Field of Mechanism and Machine Theory (2018–2022)*, Mechanisms and Machine Science 128,
https://doi.org/10.1007/978-3-031-25730-8_20

objectives [1]. In this context, this article details a teaching experience carried out for distance teaching of the design and calculation of gear transmissions.

Power transmission systems are subjected, under operating conditions, to different kinds of loads. Their failure is studied, from a theoretical point of view, by static load or fatigue, in Machine Technology at the Technical School of Industrial Engineers of the National University of Distance Education (UNED). The contents of this subject have been selected to complete the training of the engineer in the basic areas of Mechanics and Materials, and to establish the foundations that serve as a basis for more applied subjects.

The National University of Distance Education (UNED) is the Spanish public university with the largest number of enrolments, totalling more than 200,000 students, who study official degrees in almost all fields of knowledge. The main characteristic of this university is its distance-learning methodology, for which the UNED uses a virtual educational institutional platform through which the majority of the courses and subjects taught at the university are delivered and supported.

The average student profile is that of a person whose age is 37 years old and who in 80% of cases has work or family obligations. Foreign students represent 16% of the total student body, mostly from Latin America and the European Union. The economic crisis has changed the profile of the UNED student: in addition to people of a certain age who want to combine studies with work, increasingly, young people see a possibility at the UNED.

In this context, a practical exercise has been designed for the subject, whose main objectives are the following:

- That it is an ideal complement to the theoretical contents acquired in the teaching of the subject [2, 3], serves as support to the learning task and that its undertaking forms part of the continuous evaluation of the student.
- That the students learn the use of a computer program based on the finite element method (FEM).
- That it is adapted to the profile of the UNED student, carried out as virtual practice at home and not face-to-face, minimizing the cost and inconvenience due to travel.
- That it involves a reduced or zero cost in licenses both for the student and for the university.

20.2 Methodology

Once the objectives to be met by the practical exercise have been defined, the different software packages on the market are evaluated, analysing the benefits, prices, etc.

Among the very wide range of commercial mechanical engineering design and calculation programs, a program was sought that integrated mechanical design tools, such as shaft design and gear design, a calculation module using the finite element method, with the ability to perform fatigue calculations, and that the duration of the student license be extended to at least 1 academic year.

Finally, Autodesk Inventor [4] is selected. This software offers a set of easy-to-use tools for mechanical design, simulation, visualization, and 3D documentation, and

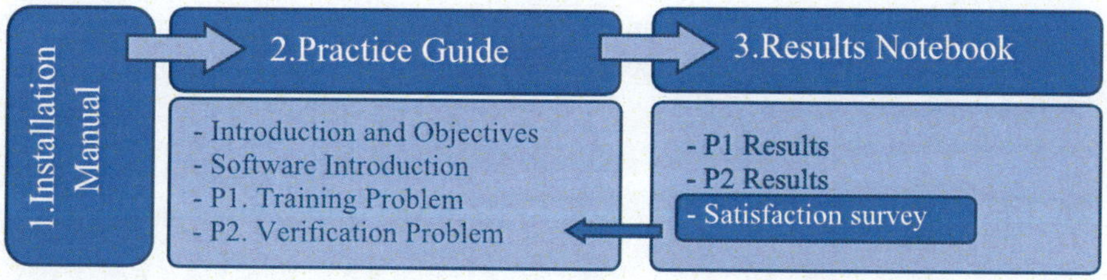

Fig. 20.1 Scheme of the information presented in the distance learning application

has free licenses for students and teachers that, in a quite uncomplicated way, allows each student to download the program on their computer.

The next step is the definition of the contents of the practice tasks. The two most important things to consider are:

- Contents: the most relevant aspects of the subject must be emphasized. The objective is to design a simple system in which the basic and fundamental principles of machine design are applied, such as the stress analysis of a power transmission shaft.
- The time that students must dedicate to undertaking the practice tasks is about 15–20 h. It is a brief time for students to learn how to use a program, so the manual must be clear, concise and with an elevated level of detail so that students learn the basic use of the program and are able to perform the tasks planned autonomously.

From the virtual educational platform, students will be able to download the three documents necessary to undertake the practice tasks (Fig. 20.1):

1. The Installation Manual is a script that explains step-by-step the procedure to download and install the software.
2. The Practice Guide, which consists of the sections:
   a. Introduction and objectives.
   b. General concepts of Autodesk Inventor.
   c. A training problem.
   d. A verification problem.
3. The Results Notebook, where the results of each one of the tasks indicated in the script for each of the problems will be written down.

20.3 Results

In this section, the student will learn, with the program, to create the geometry to be analysed, to mesh the system, to divide it into elements and nodes, to apply the boundary conditions and to obtain the unknowns or variables to be known (displacements and tensions). Each problem is divided into a series of defined tasks to guide the student through each step, enabling them to evaluate it and learn what the typical inconveniences are that usually appear when facing a problem of this type.

20.3.1 Training Problem

The objective of this problem is to learn how the program works and to study the usual issues that can arise when a system is simulated using the FEM; the results obtained will be compared with the theoretical ones and any discrepancies will be studied.

To do this, a simple system will be analysed: a circular-section bar with a step-change in section and exposed to an axial load.

This section explains the functionality of the main buttons that appear on the screen: zoom, move, drag, or rotate the image, the work tree, the creation of coordinate axes, the delete option, etc.

The first milestone consists of solving the problem theoretically. The next step will be the construction of the shaft with the program tool "Shaft Component Generator" (Fig. 20.2) that allows the student:

- To position the axis with respect to global axes.
- To generate a shaft with several sections, each with its specific length and diameter. It also allows the definition of different finishes for each end of the profile, notching radii, or defining grooves.
- To conduct theoretical calculations of the forces to which the axis is subjected. For this, the program allows the definition of the material, the type and the location of the supports and loads.

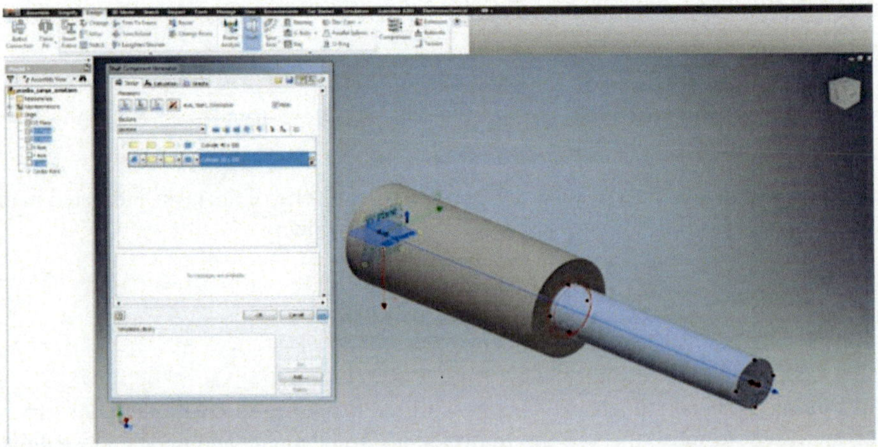

Fig. 20.2 Axis creation tool

- To visualize the diagrams of forces and moments. The comparison of these diagrams with those obtained theoretically by the student is the object of study in another of the tasks.

Once the creation of the geometry is finished, the stresses are calculated for a static analysis using the FEM. The manual explains how to create a new analysis, assign the material and define the boundary conditions (motion constraints and loads) (Fig. 20.3).

The next step explains how to mesh the system, define the main variables that influence it and how to perform the simulation. In the following tasks, the student has to identify the functionality of the main options to display the results and change the colour scales with which they are represented. In addition, the student is instructed how to place probes to visualize the results at each point.

In the following tasks, the influence of the density and the type of meshing on the precision of the final result and on the computational load required to solve the problem are studied (Fig. 20.4). The student is required to create meshes with elements of various different sizes and record the stress values, the number of nodes and elements and the time it takes for the computer to obtain the results. The student will be asked to define a finer local mesh in the zone of change of section.

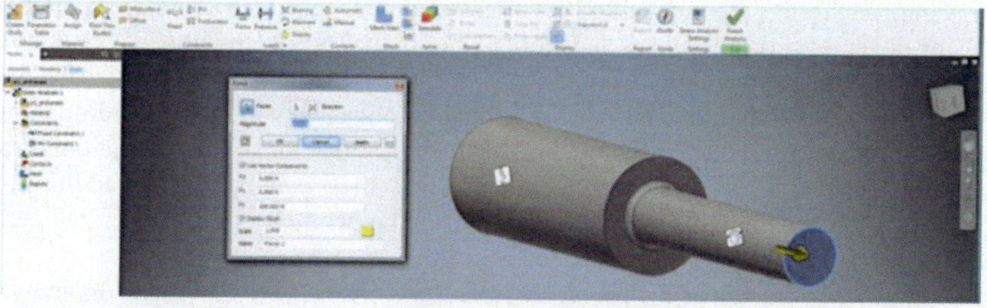

Fig. 20.3 Definition of boundary conditions

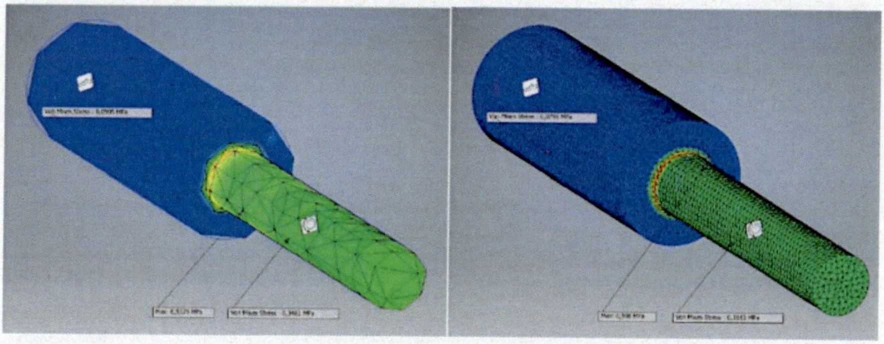

Fig. 20.4 Influence of mesh density and element type on the MEF

In addition, the calculation of the errors that the different meshes produce in the value of the tensions is required, comparing them with the theoretical values calculated.

This allows students to remember and work on:

- Stress concentration and its effect on brittle and ductile materials.
- The stress tensor and how to obtain with it the principal and the Von Mises stress.

20.3.2 Verification Problem

The level of complexity of this problem is slightly higher than the previous one and, although the fundamental steps to be followed are detailed, it is expected that the student will take the initiative for its resolution. In this case, a transmission composed of two shafts and a spur gear transmission will be analysed. Both axes consist of 5 sections of different diameters, the bearings are located at the two ends and the gear is located in the middle section. Power enters the first axle through a clutch located on the far right.

Again, the first requirement of the student is the theoretical resolution of the problem and to draw the stress and strain diagrams for each section. In Autodesk Inventor each part of the shaft must be created separately and then assembled together, using different coordinate axes and work planes in the "Shaft Component Generator" tool. A new tool "Spur Gears Component Generator" will be used to generate the pair of spur gears, which internally performs the theoretical calculations according to the chosen standard and displays them on the screen (normal, tangential, and radial force and safety factors), so that the student can verify their calculations. In addition, it simulates the meshing process in 2D and shows the geometry of the generated teeth (Fig. 20.5).

Once the geometry of the problem is finalized, a new study is created to analyse the stresses and displacements. In this case, in addition to assigning the material

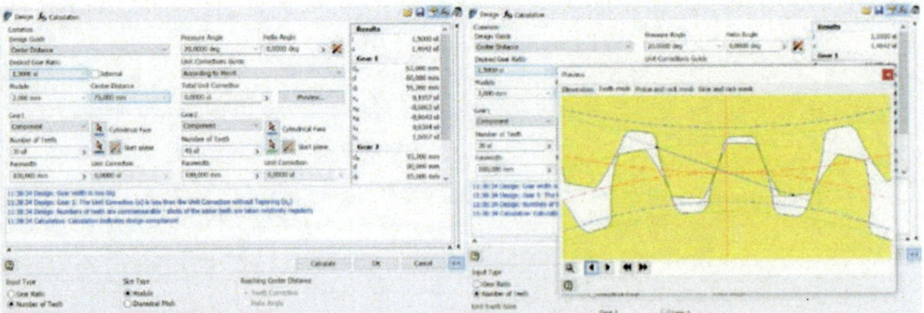

Fig. 20.5 "Spur Gears Component Generator" Tool

and establishing the boundary conditions, the contacts that exist between the distinct parts of the model must be created. An additional task for the student is to investigate the types of contacts that can be defined by the program and the difference between them: "Bonded" (between the central cylinder of the axes and the surface of the holes) and "Separation" (between the tooth of the pinion and the tooth of the wheel that are going to be in contact). The script explains how to create and define each of the contacts.

The next step is meshing. In this system it is very important to have control over the mesh as it is a larger and more complex system. A very coarse mesh can cause computational convergence problems or poor precision in the results.

On the contrary, a dense mesh throughout the model causes an extremely high computational load and requires an excessive calculation time. Therefore, the ideal solution is to establish a fairly coarse mesh in most of the model and define a much finer local mesh in the areas of interest and in the contact area of the meshing teeth (Fig. 20.6).

Once the analysis is configured, the results are calculated and obtained. In the script, the student is taught how to hide or select the elements that are interesting to visualize and analyse. Figure 20.7 shows the typical distribution of Von Mises stresses in these types of problems: the most highly-loaded areas occur at the section change, at the line of contact of the teeth in mesh, and at the base of the tooth.

The following tasks that the student must solve are related to the analysis of results and require having well-assimilated theoretical concepts. Through different probes they must identify the different components of the stress tensor and the principal stresses, obtain the diagram of the shear stress produced along the axis and compare it with the theoretical diagram.

The student must carry out a similar process to represent the stresses due to bending and analyse the possible discrepancies with the theoretical results and their

Fig. 20.6 Verification problem: boundary conditions and meshing

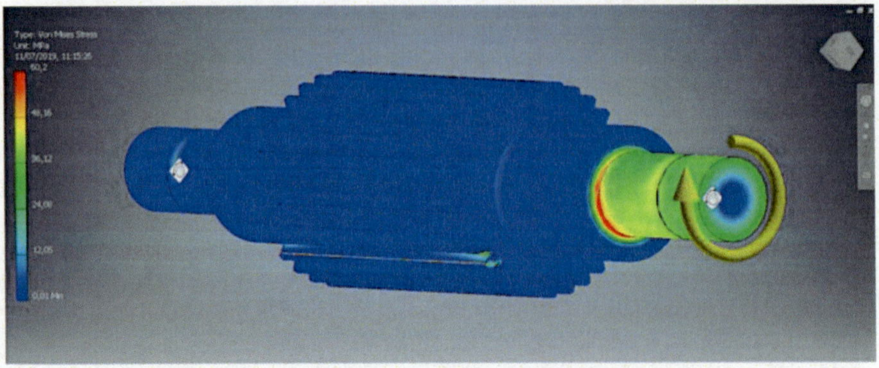

Fig. 20.7 Von Mises stresses obtained in the verification problem

causes, basically due to the boundary conditions. Figure 20.8 shows a flow chart to the basic educational path.

The last task is optional and serves to improve the given grade. The objective is to choose and justify the bearings for this model based on the simulation results: for this the student must research the website of one of the largest bearing manufacturers to choose the best possible combination of bearings.

20.4 Conclusions

From a computer tool based on the FEM, Autodesk Inventor, some practice tasks have been designed as an ideal complement to the explanations and theoretical developments carried out in the teaching of the subject, which serve as support in the learning task and are part of the continuous evaluation of the student.

A first simple problem serves to understand and learn about the program, its basic functions and to analyse the fundamental parameters that must be considered in a mechanical analysis. A second problem of greater complexity has as its objective the static analysis of the stresses obtained through simulation and their comparison with the theoretical calculations.

Through a satisfaction survey, the main problems that the student finds in the resolution of the practice tasks can be identified. This allows the analysis of possible improvements to be implemented in the scripts and in the tasks. This allows the teaching team to know if the task has aroused the student's interest and has been useful, and thus carry out a continuous improvement of the tasks.

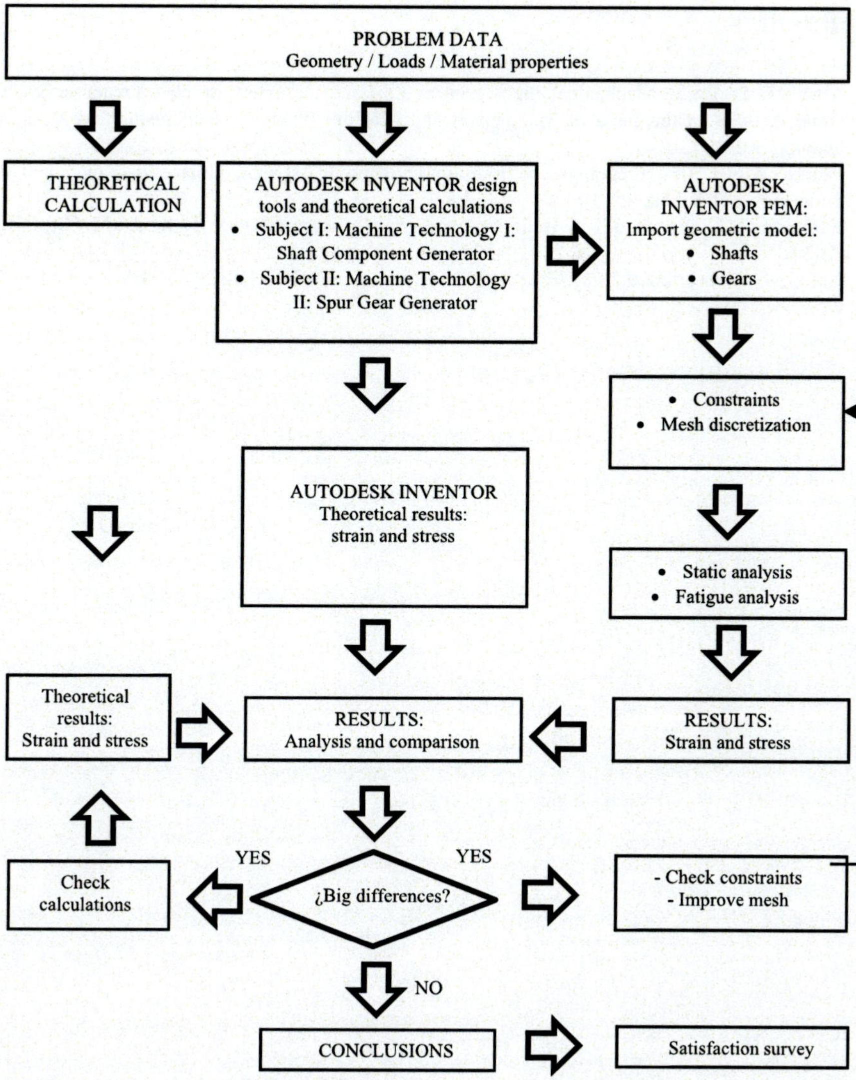

Fig. 20.8 Flow chart to the educational path

All working documents and help files cited in this article can be found in a public folder, Machine Technology Practice HELP FILES , available in the digital version of this article.

Acknowledgements This research was funded by the Spanish Council for Scientific and Technological Research, project number PID2019-110996RB-I00, and the School of Engineering of UNED, grant number 2021-MEC23.

References

1. Artés, M., López, J.: Mechanical engineering at a distance: a review. In: New Trends in Educational Activity in the Field of Mechanism and Machine Theory, Mechanisms and Machine Science, vol. 19 (2014)
2. Pedrero Moya, J.I.: Tecnología de Máquinas. Tomo I, 1st edn. Universidad Nacional de Educación a Distancia (2010)
3. Budynas, R.G., Nisbett, J.K.: Diseño en ingeniería mecánica. Shigley, 10th edn. McGraw Hill (2020)
4. Autodesk Inventor 2020. https://help.autodesk.com/view/INVNTOR/2020/ESP/

Chapter 21
Application of Reverse Engineering to Implement 3D Designs of Ancient Mechanisms

Mercedes Perdigones Gomez, Ángel Mariano Rodriguez Perez, Julio Jose Caparros Mancera, José Antonio Hernández Torres, and Cesar Antonio Rodriguez Gonzalez

Abstract In this work reverse engineering has been applied. To carry out this process, some plans and videos are used to make the mechanism in 3D. The Juanelo Turriano mechanism, which served to raise water in the city of Toledo, has been taken as a reference for the application of this work. This work is based on a video provided by the Juanelo Turriano Foundation in which the complete mechanism can be seen. To carry out the work, all the parts of which the mechanism is composed have been identified in the first place in order to be able to carry out the final assembly and be a functional mechanism. With this type of work, imagination is greatly developed since there are some pieces that have to be designed in a functional way that cannot be seen in this case in the video.

Keywords Reverse engineering · Mechanisms · Juanelo Turriano

21.1 Introduction

As for mechanisms for water supply in ancient times, the deposits in the southwest of the peninsula are notable, such as: the ten wheels of Sâo Domingos (Portugal), or

M. P. Gomez · Á. M. R. Perez · J. J. C. Mancera (✉) · J. A. H. Torres · C. A. R. Gonzalez
University of Huelva, Huelva, Spain
e-mail: Julio.caparros@diesia.uhu.es

M. P. Gomez
e-mail: Mercedes.perdigonez@alu.uhu.es

Á. M. R. Perez
e-mail: Angel.rodriguez@dci.uhu.es

J. A. H. Torres
e-mail: joseantonio.hernandez@dimme.uhu.es

C. A. R. Gonzalez
e-mail: cesar@uhu.es

J. C. García Prada et al. (eds.), *Trends in Educational Activity in the Field of Mechanism and Machine Theory (2018–2022)*, Mechanisms and Machine Science 128,
https://doi.org/10.1007/978-3-031-25730-8_21

other broken ones found in the mines of Tharsis (Huelva). Of these, not all of them are preserved, but rather they are exhibited in different museums in fractions: in the Musée des Arts et Métiers in Paris, almost 1/4 of a Sunday wheel from 1864 and 1/5 of a broken one in the Museum of Transport in Glasgow since 1867. The latter may belong to the Tharsis mines, however, its location cannot be confirmed due to the limited and confusing documentation found [1–5].

The Renaissance was a stage of great success in the evolution, development and dissemination of hydraulic machines due, in large part, to the influence of creativity that characterizes this historical period. This fact is reflected in inventions such as Leonardo Da Vinci's (1452–1519) centrifugal pump, whose dissemination throughout Western Europe in the mid-15th century is due to the invention of the printing press. In turn, the work was disclosed: "Kunstliche Abris allerhand Wasser" by Jacobus Strada in which there are different engravings such as; "hydraulic machine for spout fountain", or "Noria with draft animal" and also, the manuscript by Francisco Lobato from Medina in which he speaks of the "regolfo mill" [6, 7].

It is interesting to highlight several majestic inventions in this historical stage, however, the analysis focuses on the one that will give meaning to this document and which is discussed more extensively in the following sections. Juanelo Turriano is the creator of various machines that stood out in the European Renaissance due to their complexity and the use that was made of them. Many of his works have been recognized, while others have fallen into oblivion, such as the Artificio by Juanelo Turriano in Toledo. This majestic mechanism managed to raise the water of the Tagus River ninety meters in a journey of just three hundred and six meters. This was built for the first time in 1569 and later modified in 1581, achieving a flow of seventeen. Unfortunately, no explicit document describing this contraption has reached our days, mainly due to the creator's intention to safeguard the secrecy of the contraption's operation and numerous economic inconveniences that he found in its implementation [8–11].

21.2 Metholody

Juanelo Turriano or Janello Torriani was originally from Cremona and his date of birth was between 1500 and 1511, the earliest being the most accepted by most studies. He was born from a humble family that from an early age rubbed shoulders with illustrious characters from his environment who formed him throughout his life. Giorgio Fondulo, physicist, doctor, philosopher, connoisseur of Greek and Hebrew, mathematician and astrologer, marked Turriano's life by being his first teacher and later one of his illustrious friends [12].

The case study of this document lies in the city of Toledo and the passage of the Tagus River through it. Despite the benefit generated by this natural asset, Toledo is considered a mostly continental city, located far from the marine influence and at a medium–high elevation, being in turn shielded between its mountains, with elevations of 1000–1400 m protecting to the city of strong winds. The object of study of this work falls on the structure that revolutionized hydraulic architecture, the ingenious device of Juanelo Turriano. This section focuses on explaining the operation and purpose of this machine, whose parts will be treated in more detail in the following points. This document will explain the artifice, distinguishing it into four independent mechanisms, which, when joined together, would form the complete structure, perfectly fulfilling its objective. All the information is based on the "3D Animation of the Artifice of Juanelo Turriano" carried out by the Juanelo Turriano de Toledo association.

The mechanism had a fortress at the level of the river, where the initial drive subassemblies were located, forming a subassembly of geared wheels with water load and a subassembly of geared wheels without water load.

The first was in charge of raising the water by means of a chain provided with buckets activated by means of a reduction gear set of geared wheels. And the second, activated a tie-rod and forcing mechanism by means of an identical group of geared wheels. This ended in a wheel with a single tooth that was responsible for transforming the rectilinear circular movement of the wheels into a rectilinear oscillating one. The braces and forcing bars (Fig. 21.1) consisted of a number of bars joined together by means of pins that allowed the transmission of a rectilinear movement, overcoming irregularities in the course of the transmission of movement.

The last subset consists of a tower of oscillating buckets (Fig. 21.2) that collect water from a tank and raise it, passing it from one to another and achieving successive increases in elevation.

The motor of these towers are the braces and forcing forces that describe a horizontal rectilinear movement in the lower part, which is then transformed into another vertical rectilinear one by means of a mechanism located below. In this way, the oscillating movement of the buckets that will make the fluid rise is achieved. The braces are also responsible for connecting the different towers that will be activated in the same way as explained above.

Fig. 21.1 Tie rod and forcing mechanism

Fig. 21.2 Oscillating bucket towers

Therefore, the didactic methodology consists of analyzing the mechanism's information source, based on animation videos (as is the case), plans, sketches, diagrams, annotations, etc., in order to model a specific 3D design with capabilities scalable features. There are numerous old mechanisms, as is the case of the Juanelo mechanisms, which despite having information from various sources and nature, have not been fully adapted to new design and 3D printing technologies in educational environments. With this proposed methodology, the student starts with training in 3D design and in the fundamentals of machines and mechanisms. With this, he develops analytical skills in the functioning of mechanisms, synthesizing the key elements in functional models. Despite the complexity of the Juanelo mechanism as a whole, the choice of the Juanelo mechanism responds to its historical importance, and the large number of more simple elements that make it up, giving further study to the operation of mechanisms subdivided into different phases of operation.

21.3 Results

Based on the designs and methodology described in the previous section, a geometric, structural and mechanical analysis is proposed that results in the 3D design of the different components of the Juanelo Turriano mechanism. These are shown in detail in Table 21.1.

Figure 21.3 shows the complete mechanism once all the parts in Table 21.1 have been designed.

The bucket tower is the most important mechanism of this set since it fulfills the main objective for which this infrastructure is built, to raise the water of the river. The tower houses inside some slats with buckets at their ends that will take care of the rise of water. These perform a vertical movement through braces attached to them by pins. In turn, the straps are attached to a triangle-shaped piece at the top and another

Table 21.1 3D design of the different components of the Juanelo Turriano mechanism

Mechanism	Operation	Design
Wheel 5	Initial wheel: It is responsible for starting the transmission mechanism being driven by the flow of the river at a speed of approximately 3 m/s	
Wheel 4	Wheel 1st stage: It is connected to the initial wheel through its axis and, consequently, transmits the angular velocity of the first wheel to the next, achieve a decrease of speed thanks to the difference in the number of teeth between the two	
Wheel 3.2	2nd stage wheel: Receives the movement of wheel 4 positioned horizontally and transforms it into a vertical turning movement, thus managing to transmit this angular velocity to the next wheel through a vertical axis that connects them	
Wheel 3.1	2nd stage wheel: It is exactly the same as the wheel that precedes it and transmits the speed that arrives through the vertical cross axis to the next wheel of the mechanism	

(continued)

Table 21.1 (continued)

Mechanism	Operation	Design
Wheel 2.2	3rd stage wheel: Positioned horizontally, it receives the movement of wheel 2.2 and transmits it to the next one through a vertical cross axis that connects them	
Wheel 2.1	3rd stage wheel: It rotates with the angular speed of wheel 2.2 and transmits it to wheel 1 making its vertical cylinders collide with the teeth of the opposite	
Wheel 1	4th stage wheel: This wheel is responsible for transmitting the rotational movement to the output wheel	
Wheel 0	Output wheel: It consists of a wheel that contains a single tooth that will transmit the movement generated by the set of wheels to the bar mechanism, transforming the circular movement into rectilinear movement	

Fig. 21.3 3D model of the tie rod and forcing mechanism

of the same nature at the bottom. They will be defined more precisely in Table 21.2, where the corresponding 3D design is also shown.

Figure 21.4 shows the complete mechanism once all the parts in Table 21.2 have been designed.

21.4 Conclusions

In this work, the practical application of reverse engineering to ancient mechanisms with the use of modern digital tools has been demonstrated. In this case, applying the methodology designed to the Juanelo Turriano mechanism, which served to raise water in the city of Toledo centuries ago, a structural and mechanical analysis of the available sources has been carried out, including plans and videos of foundations on the mechanism. With this, detailed 3D designs of each of the parts of the mechanism assembly have been developed, in such a way that designs with greater capacity for dynamic analysis and capable of being able to develop scale implementation studies are proposed.

Table 21.2 3D design of the bucket tower of the Juanelo Turriano mechanism

Mechanism	Operation	Design
Connecting bar	This element fulfills the function of joining the driving wheel with the tie rod and forcing mechanism, and in turn joins the different sets of bars between them	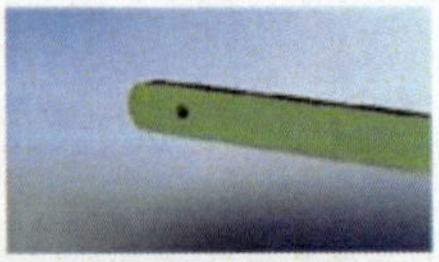
Guide bar	It consists of an extensive wooden bar that, joined laterally to a counterpart by connecting bars, transmits the oscillating movement while maintaining parallelism between them	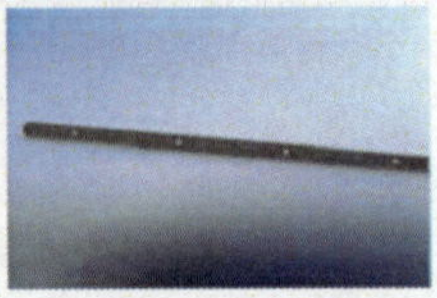
Pins	These are cylindrical pieces whose mission is to serve as a joining element for the different bars, being inserted through their holes The pin in the image is the longest in the mechanism, however, different sizes are available depending on the distance between the elements to be joined	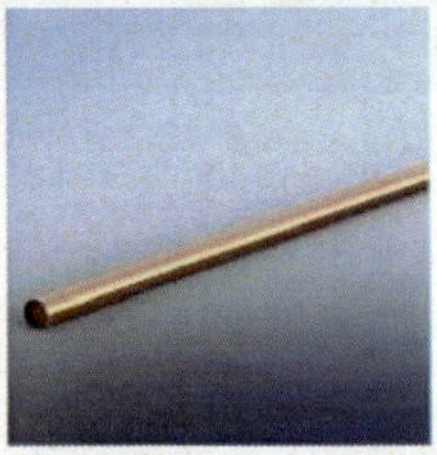
Supports	They are those that fix and limit the slenderness of the mechanism and the angle that the bars will form with respect to the ground We find two types: Supports with symmetrical reinforcements and trimmed supports. The cut supports are due to the impossibility of using the usual ones due to lack of space	 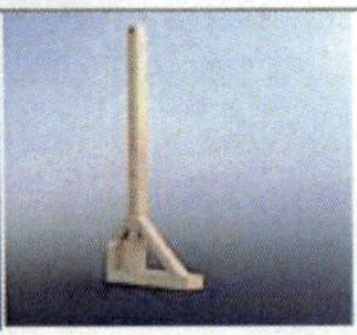
Driving part	This 1060 alloy piece is connected to the center of the tie rod and forcing mechanism by means of pins and is the motor of the bucket tower. It achieves this by dragging a piece that activates the tower when it rotates with the movement provided by the bar mechanism	

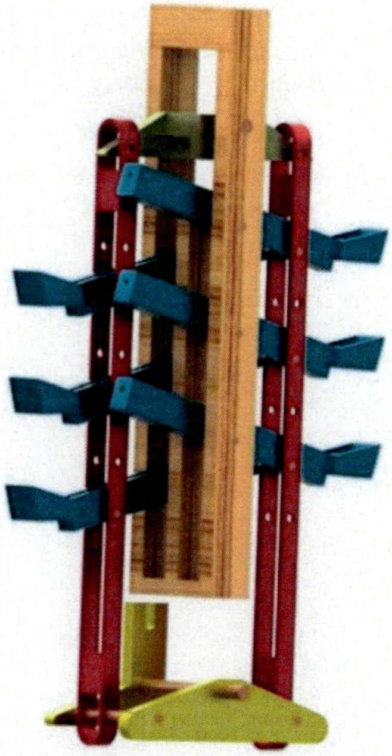

Fig. 21.4 Complete 3D design of the swinging bucket tower

References

1. Valente, T., Grande, J.A., De La Torre, M.L., Santisteban, M., Cerón, J.C.: Mineralogy and environmental relevance of AMD-precipitates from the Tharsis mines, Iberian Pyrite Belt (SW, Spain). Appl. Geochem. **39**, 11–25 (2013)
2. González, R.M., Cánovas, C.R., Olías, M., Macías, F.: Seasonal variability of extremely metal rich acid mine drainages from the Tharsis mines (SW Spain). Environ. Pollut. **259**, 113829 (2020)
3. Fernández-Caliani, J.C., Barba-Brioso, C.: Metal immobilization in hazardous contaminated minesoils after marble slurry waste application. A field assessment at the Tharsis mining district (Spain). J. Hazard. Mater. **181**(1–3), 817–826 (2010)
4. Mateus, A., Pinto, A., Alves, L.C., Matos, J.X., Figueiras, J., Neng, N.R.: Roman and modern slag at S. Domingos mine (IPB, Portugal): compositional features and implications for their long-term stability and potential reuse. Int. J. Environ. Waste Manag. **8**(1–2), 133–159 (2011)
5. Rosado, L., Morais, C., Candeias, A.E., Pinto, A.P., Guimarães, F., Mirão, J.: Weathering of S. Domingos (Iberian Pyritic Belt) abandoned mine slags. Mineral. Mag. **72**(1), 489–494 (2008)
6. Beck, T.: Jacob de Strada (etwa 1523–1588). In: Beiträge zur Geschichte des Maschinenbaues, pp. 529–537. Springer (1900)
7. García Tapia, N.: Ingeniería del agua en los códices de Leonardo y en los manuscritos españoles del siglo XVI. Ing. del agua **3**(2), 17–38 (1996)
8. Hernández-Torres, J.A., Rodríguez-Pérez, A.M., Caparrós-Mancera, J.J.: An Historical Review and Upload of Juanelo Turriano's Mechanism in Toledo, pp. 41– 48 (2022)
9. Otero, J.E., et al.: The twenty-one books of devices and machines: an encyclopedia of machines and mechanisms of the 16th century. In: International Symposium on History of Machines and Mechanisms, pp. 115–132. Springer Netherlands, Dordrecht
10. Ceccarelli, M., Cigola, M.: Trends in the drawing of mechanisms since the early Middle Ages. Proc. Inst. Mech. Eng. Part C J. Mech. Eng. Sci. **215**(3), 269–289 (2001).

11. Ceccarelli, M., Cigola, M.: Machine Designs and Drawings in Renaissance Editions of de Architectura by Marcus Vitruvius Pollio, pp. 291–307 (2016)
12. Bautista, E., Muñoz, J.L., Echávarri, J.: Juanelo (1501–1585), pp. 95–110 (2009)

Part IV
Mechanism and Machine Science Experiences in Pandemic Times

Chapter 22
Has the Teaching Innovation Carried Out During the Pandemic Been Consolidated? DEM-UPC Subjects

Lluïsa Jordi Nebot, Joan Puig-Ortiz, and Rosa Pàmies-Vilà

Abstract The alert situation due to Covid-19 caused traditional university teaching to go from face-to-face to on-line in less than a week. This article presents the experience and the measures adopted in this change of type of teaching in some subjects of the Barcelona School of Industrial Engineering of the UPC that depend on the Department of Mechanical Engineering (DEM). The subjects adopted different forms to face the new situation; those analyzed also made efforts to ensure quality teaching by developing new materials and adapting teaching and assessment systems. In some cases, it was decided to promote activities aimed at developing skills, emphasizing student self-management and, ultimately, promoting their active learning. Another aspect of this work is that it analyzes whether, really, the effort made during confinement has represented a teaching innovation in these subjects or has been just a mirage. For this, the results of two surveys carried out among the professors of the department in which they are asked about the tools they used, and still use, and their interest in sharing the experiences are presented. These results indicate that the resources used have served to improve and innovate in some aspects of DEM teaching.

Keywords Didactic resources · Teaching innovation · Moodle · GSuite

22.1 Introduction

On March 13, 2020, in accordance with the extraordinary measures due to the health emergency declared by COVID-19, face-to-face teaching activity was suspended at

L. Jordi Nebot (✉) · J. Puig-Ortiz · R. Pàmies-Vilà
Mechanical Engineering Department, Universitat Politècnica de Catalunya, Barcelona, Spain
e-mail: lluisa.jordi@upc.edu

J. Puig-Ortiz
e-mail: joan.puig@upc.edu

R. Pàmies-Vilà
e-mail: rosa.pamies@upc.edu

J. C. García Prada et al. (eds.), *Trends in Educational Activity in the Field of Mechanism and Machine Theory (2018–2022)*, Mechanisms and Machine Science 128,
https://doi.org/10.1007/978-3-031-25730-8_22

the Barcelona School of Industrial Engineering (ETSEIB) as in the rest of the country. This closure impacted 90% of the world student population, according to data from the United Nations Educational, Scientific and Cultural Organization [1].

The management teams of the different schools urged to follow the academic course in a non-face-to-face way, making use of the virtual campuses. This required stoppage forced all teaching teams to reinvent themselves in order to continue classes remotely. The urgency with which the pandemic was unleashed made it go from face-to-face classes on the mainland of the classroom, with the support of blackboards, transparencies, models… to non-face-to-face classes, with the private media that each teacher could have in their home, in a short period of five working days.

The Moodle platform (Atenea, in the case of the UPC) became the backbone of the organization and Google's GSuite became the communication tool with the students. The Institute of Education Sciences (ICE) of the UPC, as well as members of the teaching and research staff at the individual level, subsequently offered a wide variety of courses and seminars to train teachers to carry out non-face-to-face teaching and assessment.

The cases presented correspond to two subjects of the bachelor's degree in Industrial Technology Engineering (grau en Enginyeria en Tecnologies Industrials, GETI) of the ETSEIB. Specifically, they are Mechanics from the 3rd semester and Machine and Mechanism Theory from the 4th semester of the GETI. Both are compulsory subjects with a high number of students and the authors of this paper have been teachers of both subjects.

In addition, in this work, the authors, awarded for some teaching innovation actions, wish to present the results of the questions that were raised about whether the use of the teaching resources used during the pandemic really represented an innovation or if they were only an emergency resource to overcome the confinement situation. Obviously, this is a topic that other authors also raise [2, 3].

The authors promoted a day of exchange of experiences between all the members of the department and of which, in this article, the results are presented.

22.2 Teaching During the Pandemic

The teaching staff of the subjects dealt with in this paper maintained permanent communication and collaboration throughout the teaching and the non-face-to-face evaluation process, as well as with other professors of the department.

22.2.1 Mechanics (3rd Semester GETI)

The Mechanics teaching team decided that the students had to be protagonists of the learning process and feel committed to actively participating in it [4]. During confinement, students had a more active and autonomous role, acquiring skills and building their own learning process. At the same time, teachers became guides in the learning process, generating and providing new, diverse and extensive material to adapt to the abilities of each student and to help them acquire the skills proposed in the subject.

Teachers thought that assessment could be the common thread that kept students motivated and engaged.

The evaluation, ideally, has to be formative, continuous and competency-based [5], in addition to being transparent and shared from the beginning. The virtual environment is an ideal tool to obtain this visibility and accessibility at all times. In addition, facilitating self-assessment mechanisms is a highly recommended proposal in a subject with a large number of students such as Mechanics [6].

Finally, teaching was based on the creation of PDF documents (creation of new material with basic theoretical content to facilitate learning objectives), videos and animations, discussion forums (to promote self-learning and encourage collaborative work) and live sessions through the GSuite Meet, which were recorded and made available to students for viewing.

The Virtual Campus was organized by teaching units where the necessary content and a self-assessment (in a test format of 4 questions from a wide bank of questions) were posted weekly. In addition, each unit had a continuous assessment exam (synchronous online) that allowed the student to take the subject up to date.

One of the notable results is that the material created corresponds to a very broad bank of questions; the preparation of the self-assessment questions and the exam tests generated 728 questions and a collection of 62 solved exercises [7].

22.2.2 Machine and Mechanism Theory (4th Semester GETI)

The teaching team of Machine and Mechanism Theory decided that the content of the subject would not change and that they would continue teaching with the same schedule and the same groups as in the face-to-face classes, with some slight modifications. The face-to-face classes lasted 1 h 40 min [8].

For non-face-to-face classes, it was decided to make videos, mostly with the Screencast-O-Matic tool [9] in its free version, with theoretical content and problems of a maximum duration of 15 min, which the students had to visualize, work and understand during the first part of the class. This duration is considered adequate since they represent small "pills" of information and do not become heavy in their visualization. From the virtual campus of the subject, it was possible to access the videos posted on the YouTube platform. The second part of the class would be devoted to solving students' doubts through Meet sessions with the live participation of teachers and students. In this way, the non-face-to-face classes would resemble the face-to-face classes in the classroom as much as possible. It was decided to keep the evaluation system.

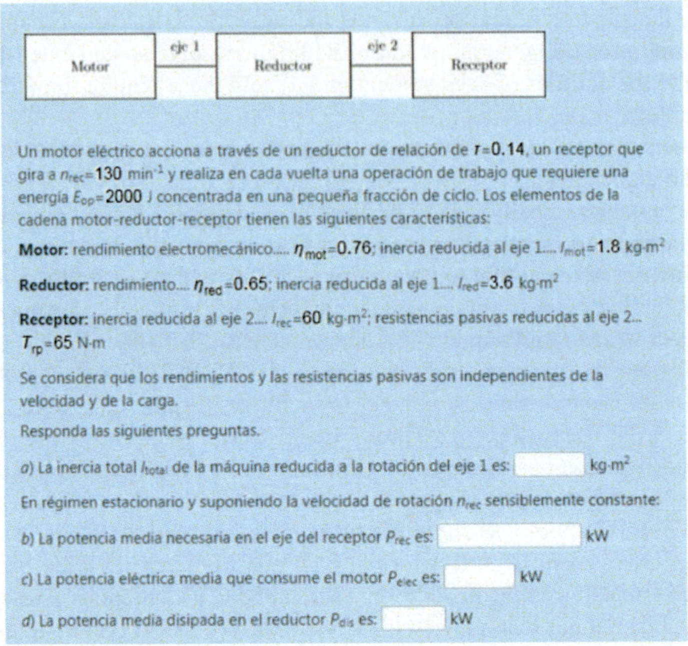

Fig. 22.1 Numerical response exercise with random variation of parameters and automatic correction in the visualization version [8]

As an example, the final exam, which lasted 3 h, consisted of the synchronous resolution of three problems, as it would have been if the exam had been face-to-face. The difference was that the problems had to be solved sequentially. Of the three problems, two were of numerical answers with the same statement for all students, but with random variation of parameters and automatic correction. They were prepared using the WirisQuizzes tool [10]. Figure 22.1 shows an example of this type of exercise. The third problem, with parameter variation, was corrected manually. The students had to submit it in PDF format on the digital campus of the subject.

One of the notable results is that a total of 120 videos were made, 60 in Spanish and 60 in English, which continue being used as reinforcement for classes since they are available to students on the virtual campus [8].

22.3 Teaching After the Pandemic

22.3.1 Year 2020–2021

The initial approach, in which classes could be face-to-face, disappeared in the second half of October. Classes were held remotely again, although face-to-face practices

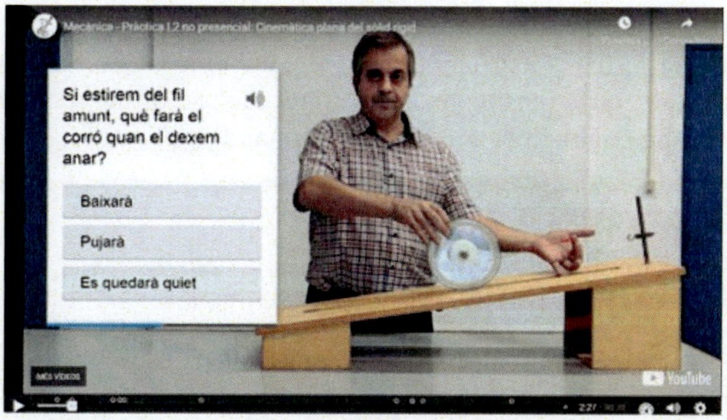

Fig. 22.2 Example of lab practice video with H5p layer [7]

were allowed. Thus, the use of the teaching tools used during the pandemic continued and some of them were improved.

Both subjects raised the theory classes and problems in the same way: retransmission of the classes in streaming through Google Meet.

Mechanics decided that the content of the practices would be carried out through videos enriched with an H5p layer (Fig. 22.2) while Machine and Mechanism Theory decided to carry them out in person.

22.3.2 Year 2021–2022

The academic year was completely face-to-face, so some of the resources used during the pandemic and the following year lost their reason for being.

However, the teaching material created continued to be available to students through the digital campus of the subjects.

22.4 Teaching Improvement

In order to analyze whether the efforts made during the pandemic and the year after have represented an improvement and an innovation in the teaching of the different groups of the UPC DEM, the authors proposed two surveys among their department colleagues. This is a topic that other authors also raise [2, 3, 11, 12].

It must be taken into account that the DEM teaches, often on the same subject, in five different UPC schools, each with its own degrees and study plans. Thus, it is common for the members of a center to be unaware of the teaching tools or resources used by colleagues in other centers.

In the first survey, questions have been asked about the tools used, the difficulties encountered and the interest in participating in a working session to share experiences. In the second one, questions were asked about the use of the tools in four periods: before and during confinement, after the return and now.

22.4.1 Survey on Teaching Experiences in the Pandemic

This survey was conducted during the second half of July 2021 and was prepared using Google Forms. The questions were:

- What tools or resources are you using and do you think they help improve your teaching practice? What virtual campus options are working for you? Do you use programs like Mentimeter, Padlet, Socrative, Jamboard, *Kahoot!*…?
- What are the problems you are having or what would you like to improve and maybe you don't know the resources that allow you to do it?
- In which subjects have you applied these resources?
- Do you think that any of your experiences can be extrapolated to other subjects and would you like to share them?
- Would you like to participate in a meeting to learn about new tools?

As expected, the tools used by teachers are very diverse and each one uses those with which they feel most comfortable and familiar. The most notable problems that arose were the interaction with the students and the possibility of obtaining feedback more quickly. As regards the interest in participating in a day of exchange of experiences, the response was unanimously positive.

22.4.2 Survey on Teaching Tools Applied to Mechanical Engineering

Based on the responses to the first survey, the authors analyzed how to address the issues raised. It was decided to carry out a second survey based on the use of specific tools, extracted from the first survey, in order to focus the teaching day of the department.

It was carried out during the second half of March 2022 and was prepared using the Jotform tool. It asked about 35 tools grouped into three blocks: Moodle resources, Google resources and others. Only two questions were asked whose possible answers appeared in a drop-down menu:

- Usefulness of the tool in four periods: **before** the pandemic, **during** the first confinement (2019–2020 academic year, spring semester), after **the return** (2020–2021 academic year) and **today** (2021–2022 academic year). In each case, the possible answers and their score were: I don't use it (0); nothing useful (2.5); not very useful (5); quite useful (7.5); very useful (10).

- Evolution of the use of the tool from the pandemic to the present. The possible answers were: I have never used it; I think it is only suitable for online teaching; I think it is only suitable for face-to-face teaching; I discovered the tool during the pandemic and continue to use it; I think it's useful but I don't use it anymore; I have used the tool in both online and face-to-face teaching; others: explain it in the final comment, please.

The answers to the first question allow us to define a utility index, *use* (Eq. 22.1), as a weighted average:

$$use = \frac{\sum_i num.\ answers_i \cdot p_i}{10\sum num.\ answers} \tag{22.1}$$

where p_i is the score given to the response i and *num. answers*$_i$ the number of people who have given that rating.

The results show (Fig. 22.3) that the use of the different tools reached its peak during confinement. Currently, although their use has decreased, these tools are used more than in the pre-pandemic period [13].

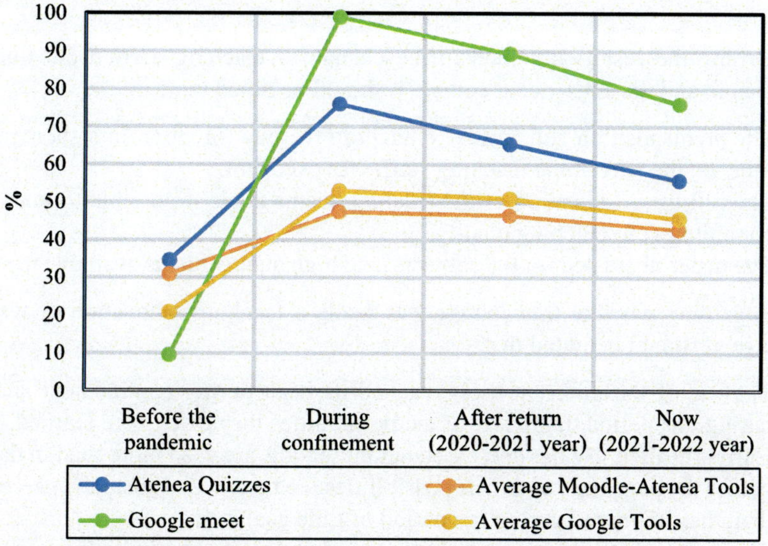

Fig. 22.3 Utility index for the Moodle-Atenea tools and Google tools. The utility index of the most outstanding tool of each group is also shown

22.5 Teaching Day of the DEM

The conference was held on April 27, 2022, lasted 3 h and was carried out in a non-face-to-face format using the Google Meet tool. The conference was offered through the ICE-UPC, open to UPC faculty and in particular addressed to the Department. There were six members of the DEM, apart from the authors, who acted as speakers and had the participation of another 18 members of the Department. Both the attendees and the speakers obtained the certificate of participation and teaching as an ICE training activity.

This made it easier to have an objective assessment of the day since after all the ICE activities, the participants receive a satisfaction survey whose questions are [13]:

1. My assessment of the academic aspects of the activity is positive: 4.88/5.
2. My level of satisfaction with the teacher is positive: 4.79/5.
3. The general organization of the activity has been adequate: 4.94/5.

The topics covered in the conference were: Edpuzzle, Overleaf, WirisQuizzes, Padlet, Socrative, *Kahoot!*, H5p, Augmented Class and Atenea.

22.6 Conclusions

Regarding the results obtained in the change of teaching carried out during the pandemic and the subsequent course, it should be noted that:

- The predisposition and ability to adapt to the new situation in a short period of time, on the part of the teaching staff, was excellent.
- The full-time dedication of the teaching teams made it possible to maintain the commitment to teaching quality.
- The material created for the subjects is of high quality and is available to students.

Regarding whether this change was really a teaching innovation or was just a mirage, it should be noted that:

- The use of the different tools reached its peak during confinement and it was during this period that many of the possibilities they offer were learned.
- Currently, their use has decreased but these tools are used more than in the period before the pandemic and their possibilities are better used since the perception of them has improved during the period of time analyzed.
- The resources used have served to improve and innovate in some aspects of DEM teaching.

References

1. UNESCO: Crisis y currículo durante el COVID-19: Mantención de los resultados de calidad en el contexto del aprendizaje remoto, Respuesta del ámbito educativo de la UNESCO al COVID-19 (2020). https://unesdoc.unesco.org/ark:/48223/pf0000373273_spa
2. Crawford, J., et al.: COVID-19: 20 countries' higher education intra-period digital pedagogy responses. J. Appl. Learn. Teach. **3**(1), 9–28 (2020)
3. Eder, R.: The remoteness of remote learning. J. Interdisc. Stud. Educ. **9**(1), 168–171 (2020)
4. Gibbs, G., Habeshaw, T.: Preparing to teach: an introduction to effective teaching in higher education. Technical and Educational Services. Bristol (1989)
5. Kaftan, J.M., Buch, G.A., Haack, A.: Using formative assessment to individualize instruction and promote learning. Middle Sch. J. **37**(4), 44–49 (2006)
6. Huisman, B., Saab, N., Van Driel, J., Van den Broek, P.: Peer-feedback on academic writing: undergraduate students' peer-feedback role, peer-feedback perceptions and essay performance. Assess & Eval High Educ **43**(6), 955–968 (2008)
7. Pàmies-Vilà, R., Clos Costa, D., Bosch Soldevila, E., Barjau Condomines, A.: Docencia no presencial de emergencia en materias troncales con número elevado de estudiantes. Caso práctico: asignatura Mecánica del Grado en Ingeniería en Tecnologías Industriales. In: Proceedings of XXIII Congreso Nacional de Ingeniería Mecánica, Jaén, 2021 (2021)
8. Jordi Nebot, L., Puig-Ortiz, J., Zayas Figueras, E., Pérez Gracia, A.: De tierra firme a la nube en un salto: Teoría de Máquinas y Mecanismos y Vibraciones Mecánicas en la ETSEIB. In: Proceedings of XXIII Congreso Nacional de Ingeniería Mecánica, Jaén, 2021 (2021)
9. https://screencast-o-matic.com. Accessed 19 Sept 2022
10. http://www.wiris.com/es/quizzes. Accessed 19 Sept 2022
11. Elfirdoussi, S., et al.: Assessing distance learning in higher education during the COVID-19 pandemic. Educ. Res. Int. (2020). https://doi.org/10.1155/2020/8890633
12. Mishra, L., Gupta, T., Shree, A.: Online teaching-learning in higher education during lockdown period of COVID-19 pandemic. Int. J. Educ. Res. Open (2020). https://doi.org/10.1016/j.ijedro.2020.100012
13. Jordi Nebot L, Puig-Ortiz J, Pàmies Vilà R (2022) Recursos didácticos de pandemia: ¿innovación docente o espejismo? In proceedings of XV Congreso Iberoamericano de Ingeniería Mecánica. Madrid, 2022. Pending publication

Chapter 23
Implementation of Audiovisual Material in Lab Sessions During COVID Time: Effects and Results

Raúl Gismeros Moreno, Eduardo Corral Abad, Abraham Vadillo Morillas, and Cristina Castejón Sisamón

Abstract The COVID-19 pandemic was a turning point in how teaching activity was carried out. The sudden lockdowns and the subsequent health restrictions made the teachers improvise and adapt their approach in a short time, making use of new technologies. This had a special impact on the classes developed in laboratories. In the case of this work, this imposed change was also the perfect time to address different issues that had hampered the grades obtained by the students and distorted the spirit of "lab classes". The subject of this study, *Machine Theory*, has proved to respond positively when implementing elements based on audiovisual technologies. As can be seen throughout this paper, the results show that the grades have improved, the health regulations have been fulfilled and the students are satisfied with the new approach.

Keywords Laboratory Classes · Machine Theory · European Higher Education Area (EHEA) · Teaching

23.1 Introduction

The COVID-19 pandemic that shook up the world and whose effects are even felt nowadays had also an impact on teaching activity [1]. Faculty teams were forced to adapt, in a short time, to a remote, online format, in order to keep the academic year going on. This shift affected especially laboratory teaching [2, 3]. Concepts such as e-learning or virtual labs arose as alternatives to overcome the first restrictions [4]. However, once the main lockdown was lifted, a second challenge came up: keeping the new situation for the time being or going back to the previous reality. The own philosophy of the Bologna Process encourages a student-centred learning process with special emphasis on both experimental and team-focused work [5, 6]. These

R. Gismeros Moreno (✉) · E. Corral Abad · A. Vadillo Morillas · C. Castejón Sisamón
MaqLab Research Group, Department of Mechanical Engineering, Universidad Carlos III de Madrid, Avda. de La Universidad, nº30, 28911 Leganés, Spain
e-mail: rgismero@ing.uc3m.es

J. C. García Prada et al. (eds.), *Trends in Educational Activity in the Field of Mechanism and Machine Theory (2018–2022)*, Mechanisms and Machine Science 128,
https://doi.org/10.1007/978-3-031-25730-8_23

two features would be greatly damaged if the former approach was kept. For this reason, different alternatives were proposed to keep the in-presence aspect of lab sessions, but, at the same time, fulfilling the social distance requirements defined by health authorities [7, 8].

This work focuses on one specific subject, *Machine Theory*, which is taught in the third year of the Bachelor of Mechanical Engineering at University Carlos III de Madrid (UC3M). This course distinguishes itself for responding well to the implementation of new technologies to ease the understanding of the theoretical content of the subject [9, 10], so it is a good testing ground to try alternative methodologies. Laboratory sessions make up 20% of the final grade, the rest of it divided between another 20% related to the mid-term exams and 60% of the final exam. The two points of the final grade associated with the laboratory classes are split up equally into simple tests of the content of each session and the reports handed in by the students after the sessions. They organise themselves into groups of four or five people, having four or five groups per class (20–25 students in total). Four laboratory sessions are developed throughout the semester, three of them are related to different themes of the subject (*machines and introduction to synthesis*, *cams* and *gears*), whereas the last class consists of a seminar hosted by a guest from renowned engineering companies such as SKF or NTN-SNR to show the students alternative perspectives. This last session is not considered in the final grade but is mandatory for all students.

The rest of the work is structured as follows: in Sect. 23.2, the approach to laboratory sessions before the pandemic is described, underlining the main problems identified. Then, the new methodology implemented is developed in Sect. 23.3. Finally, an analysis of the effects associated with this change is carried out in Sect. 23.4.

23.2 Pre-pandemic Approach

During the courses before the lockdown, laboratory sessions were divided into two parts: a 10-min test on arrival to make the students read the script provided beforehand and the experimental work in the lab. The students should hand in a brief report when leaving the class, so no homework was made. This way of doing had two consequences: more than half of the students didn't read the scripts before the classes, as they knew that someone on their teams would do the work, which resulted in unpleasant situations as usual cheating during these tests (half of the grade of each session was associated to them); and the quality of the reports was poor (see Fig. 23.1) since the students only had about two one hour and a half to complete the experimental work and discuss the results obtained. Because of the lack of preliminary work before the session, teachers were forced to explain concepts already developed in the scripts and the theory classes, which also contributed to the bad quality of the reports. This also went against the spirit of the Bologna Process, as what was supposed to be student-centred work turned into teacher-guided exercises, with almost no difference from the problem sessions carried out in normal classes.

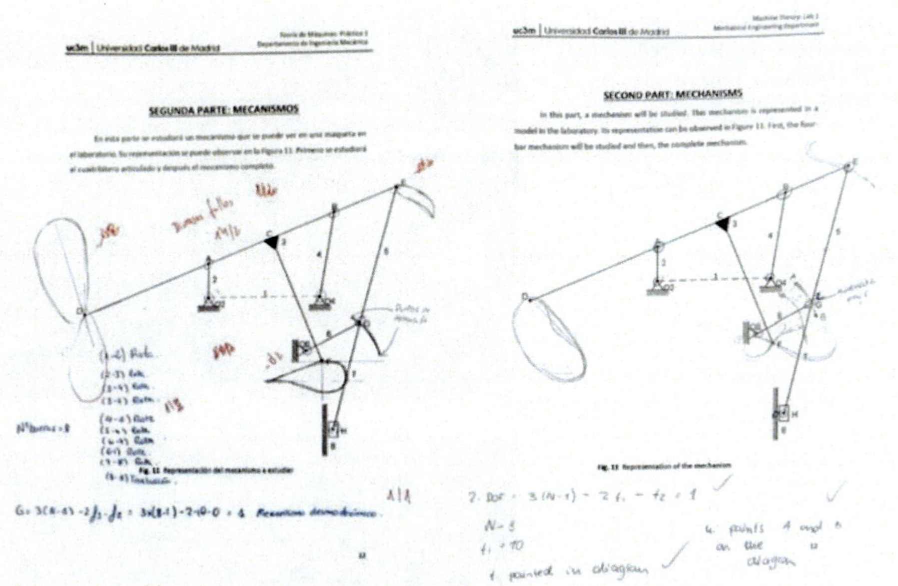

(a) (b)

Fig. 23.1 Example of one of the reports handed in by the students, with the teacher's annotations highlighted in red: **a** Spanish version; **b** English version

23.3 Post-pandemic Approach

As the aforementioned subject is developed during the first semester of the academic year, it was not affected by the toughest restrictions established during the lockdown in the 2019–2020 course. However, the social distancing regulations defined for the 2020–2021 program established a maximum number of people per class, which made it impossible to have 25 people in a space like the one shown in Fig. 23.2, with dimensions 7.82 × 4.82 m. Health recommendations stated that exposure time should be reduced as much as possible to prevent the expansion of the virus. In order to keep the "experimental experience" for all students and obey the health regulations, it was decided that there would be more groups of students (up to six) per class, with three people per group. For each session, only one member of each group would attend the laboratory and would take the data required to complete the report, which would not be handed in when leaving the class, but within a one-week deadline. This change brought a positive effect in the form of an improvement in the quality of the reports, as students had more time to work on them. Apart from this, on-arrival tests were removed and replaced by a single test that was made along with one of the midterm exams that were done throughout the semester. The one-member-per-group restriction was only applied for the 2020–2021 course.

In order to improve the "quality time" of the session and ensure that students knew what the class was about, introductory videos were developed. The contents of

Fig. 23.2 Laboratory where lab sessions are developed. The windows were installed before summer this year

these were the same as those covered by the scripts, but the main difference was the inclusion of elements that could support the exposition and show the practical applications of the theoretical part. For example, for the first session, the diagram shown in Fig. 23.3a, which describes the different systems and subsystems of a machine, was introduced to the students, with the sole support of the components of the lab. Conversely, in the video, whose captures are included in Fig. 23.3b, the students can see real examples of the described systems. As can be observed, the videos included subtitles, so students can watch them wherever they want, for example, while commuting to and from the university, without the problem of requiring headphones or being in a quiet place. Another example of the enhancement of the explanations is shown in Fig. 23.4. This video corresponds to the second session, where the students are required to draw the profiles of a cam with a roller follower. The drawing process is detailed in the script, but there are always questions in class about it and teachers end up doing it live. As the manipulation of the drawing material of different students can contribute to the spread of the virus, it was decided to describe this process with a CAD software while including the text of the script. Students

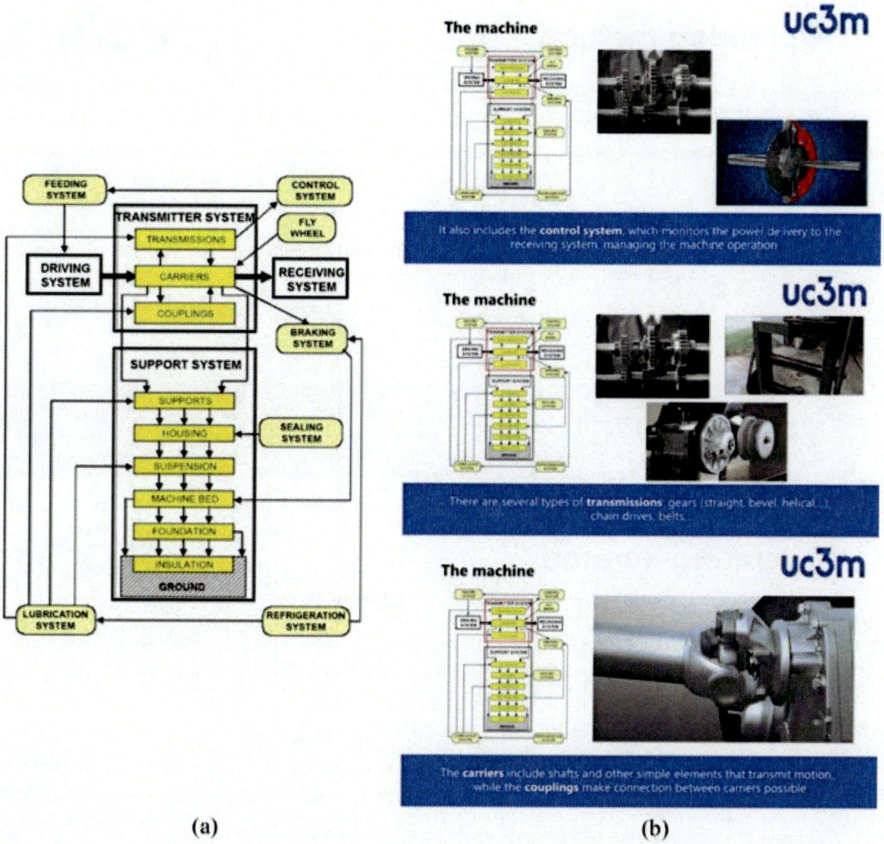

Fig. 23.3 Several captures of the introductory video of the first session: **a** diagram used; **b** examples of audiovisual material used in the video

appreciated it, as it was also helpful for the theoretical classes, and the drawings in the reports improved substantially.

23.4 Impact of the Approach Change on the Academic Results

An analysis of the evolution of the lab grades in the last four courses has been carried out. The values of the grades, both for the reports and the tests, are shown in Fig. 23.5. Spanish and English groups have been considered independently, as the students that compose them are from different backgrounds (except for the first post-pandemic course, when there weren't that many exchange students). The first two courses were

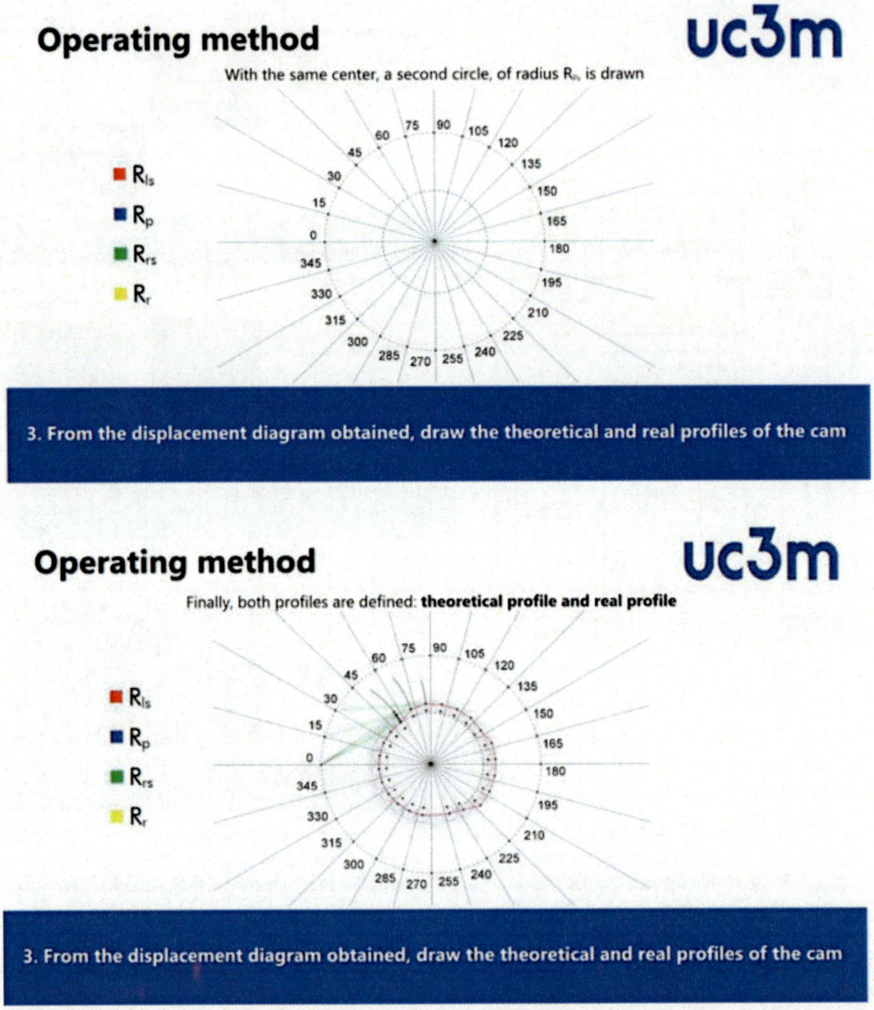

Fig. 23.4 Several captures of the introductory video of the second session, where the drawing process of the profiles of a cam with a roller follower is described

done under the former methodology, whereas the last two have been developed with the approach described above.

As can be seen, there is a slight improvement in the grades, for both groups. This improvement is not representative in the English group during the 20–21 course, as the composition of this group was different from that of previous and following years and there was a restriction of one member per group. However, the betterment of the quality of the reports is clear, as the evaluation criteria have been greatly tightened since students have now more time to work on them, but the grades have kept constant or even improved. For the tests, the 18–19 Spanish group grade could

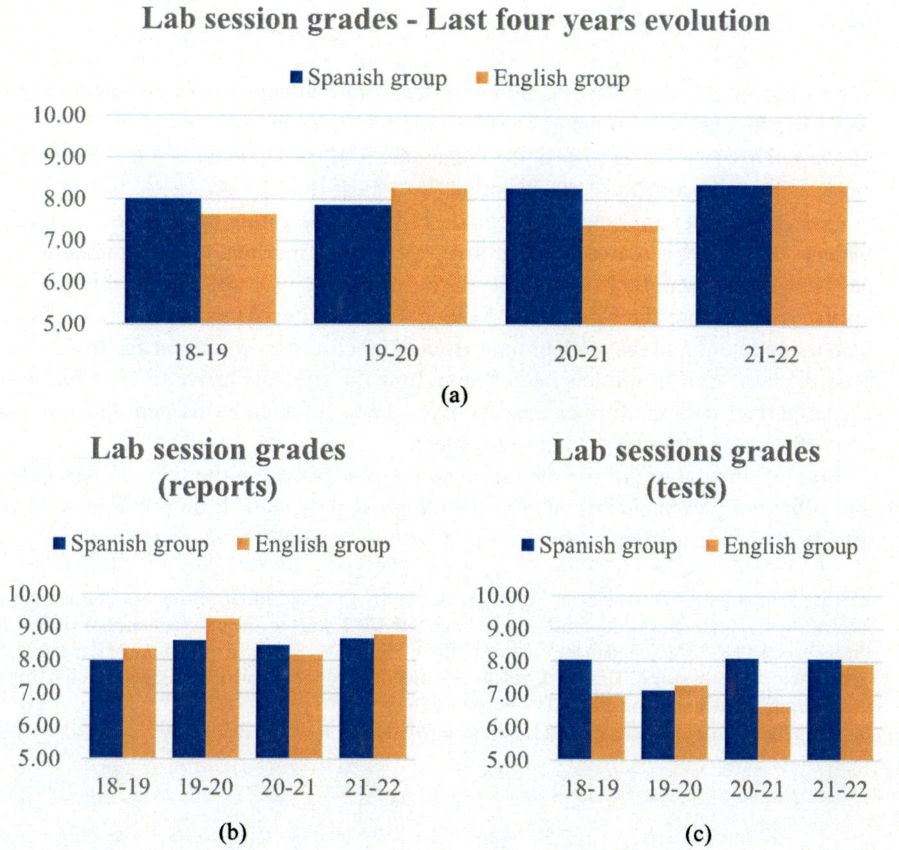

Fig. 23.5 Evolution of the grades related to the laboratory classes, for the last four courses

be taken with a pinch of salt, as there were several cases of cheating that could have disturbed the value. Nevertheless, the increase is clear for both post-pandemic courses. It was noted that students massively made use of the videos to prepare for this sole test done almost at the end of the semester. Another indicator that validates the improvement of these sessions is the increase in the students' satisfaction, reflected in several acknowledgements from the vice-rectorate for the teaching staff of UC3M after analysing the surveys completed by students about their teachers at the end of each semester.

23.5 Concluding Remarks

The extraordinary situation created by the COVID-19 pandemic forced professors to perform some necessary changes to the approach to the lab sessions on the subject of *Machine Theory* in order to ease the engagement of the students in the experimental work, within the context of the Bologna Process of the EHEA.

First, the former approach has been described, underlining the main issues identified and the consequences of the health restrictions on it. Then, the new methodology has been introduced, focusing on the effects of the inclusion of audiovisual material on the work before the session. The video-supported explanation is more efficient than a document and the student understands much faster, which means less wasted time in class. And last, it has been shown how the students arrive to the class more prepared than with the former approach. Students get another element that can also be useful in the other aspects of the subject.

Finally, an analysis of the evolution of the grades during the two courses before and after the pandemic has shown that there is a noticeable improvement in the results, the students being more satisfied with the running of these sessions.

Acknowledgements This work has been supported by the Madrid Government (Comunidad de Madrid-Spain) under the Multiannual Agreement with UC3M in the line of Excellence of University Professors (EPUC3M20), and in the context of the V PRICIT (Regional Programme of Research and Technological Innovation). This work has also received the support from the European Union trough its program NextGenerationEU/PRTR, as part of the project TED2021-131372A-I00, managed by the Spanish Ministry of Science and Innovation (MCIN/AEI/10.13039/501100011033).

References

1. Kumar, A., et al.: Impact of the COVID-19 pandemic on teaching and learning in health professional education: a mixed methods study protocol. BMC Med. Educ. **21**(1), 439 (2021). https://doi.org/10.1186/s12909-021-02871-w
2. Ray, S., Srivastava, S.: Virtualization of science education: a lesson from the COVID-19 pandemic. J. Proteins Proteomics **11**(2), 77–80 (2020). https://doi.org/10.1007/s42485-020-00038-7
3. Díez-Pascual, A.M., Jurado-Sánchez, B.: Remote teaching of chemistry laboratory courses during COVID-19. J. Chem. Educ. (2022). https://doi.org/10.1021/acs.jchemed.2c00022
4. Waldrop, M.M.: Education online: the virtual lab. Nature **499**(7458), 268–270 (2013). https://doi.org/10.1038/499268a
5. Hoidn, S., Klemenčič, M.: The Routledge International Handbook of Student-Centered Learning and Teaching in Higher Education (2020). https://doi.org/10.4324/9780429259371
6. Gargallo, B., Mateo, P.M.S., Verde, I., Almerich, G.: What happens when teachers use learning-centered methods? Effects on learning approaches, on students' capacities and on their perception of learning environment. Rev. Educ. **2018**(382), 163–197 (2018). https://doi.org/10.4438/1988-592X-RE-2018-382-396
7. Gasparello, J., et al.: Teaching during COVID-19 pandemic in practical laboratory classes of applied biochemistry and pharmacology: a validated fast and simple protocol for detection of SARS-CoV-2 Spike sequences. PLoS One **17**(4):e0266419 (2022). https://doi.org/10.1371/journal.pone.0266419

8. Agnelli, B., et al.: Blended practical learning in compliance with COVID-19 social distancing. SN Soc. Sci. **2**(5), 1–15 (2022). https://doi.org/10.1007/s43545-022-00358-z
9. Corral Abad, E., Gómez García, M.J., Ruiz Blázquez, R., Bustos-Caballero, A., García-Prada, J.C.: Using an android app for teaching mechanism and machine theory. Mech. Mach. Sci. **64**, 103–110 (2019). https://doi.org/10.1007/978-3-030-00108-7_12.
10. Corral Abad, E., Gómez García, M.J., Ruiz Blázquez, R., Castejon, C., García-Prada, J.C.: Effects of an android app on mechanical engineering students. Comput. Appl. Eng. Educ. **26**(4), 1050–1057 (2018). https://doi.org/10.1002/cae.21955

Chapter 24
New Formative Evaluation Methodology on Rotating Machinery Diagnostics

Marta Zamorano, María Jesús Gómez García, and Cristina Castejón Sisamón

Abstract Advanced Technologies in Analysis and Diagnostic of Machinery is taught in the Master in Industrial Mechanics at Universidad Carlos III de Madrid. The difficulty of this subject is that most of the theoretical contents are advanced, and at the same time new for the students. Besides their training is uneven and the number of teaching hours is short, which makes that sometimes the contents are not easy to consolidate. At the year 2017, it was observed that most students did not participate when a question was asked by the professor. Besides, the average final exam mark was low. For this reason, from 2018 it was decided to remove the group work, that had a strong experimental component, and avoid the real measurements on the part of the students. In its place it was introduced an individual work, that involves the diagnosis of rotating machinery using vibration signals given to the students (they do not measure them). The application of the personal work is used as example during the classes, to reinforce the theoretical contents. Until that moment, the evaluation strategy was purely summative. In 2020, a new formative evaluation strategy was introduced. It is based on online questionnaires that encourage students to keep up-to-date with the course. The questionnaires were introduced in the practical sessions, accounting for 10% of the final grade. This provided valuable information for the teacher, because it is possible to see in real time if the concepts have been consolidated or not, and rethink strategies. The implementation of the questionnaires is useful also for self-evaluation of the students, since they can detect gaps in knowledge. It has been observed that the final exam grade has notably improved with the implementation of these actions.

M. Zamorano · M. J. G. García (✉) · C. C. Sisamón
Universidad Carlos III de Madrid, Avenida de La Universidad, 30, 28911 Madrid, Spain
e-mail: mjggarci@ing.uc3m.es

M. Zamorano
e-mail: mazamora@ing.uc3m.es

C. C. Sisamón
e-mail: castejon@ing.uc3m.es

J. C. García Prada et al. (eds.), *Trends in Educational Activity in the Field of Mechanism and Machine Theory (2018–2022)*, Mechanisms and Machine Science 128,
https://doi.org/10.1007/978-3-031-25730-8_24

Keywords Online questionnaires · Machinery diagnostics · Practical work · Vibration analysis

24.1 Introduction

The subject 'Advanced Technologies in Analysis and Diagnostic of Machinery' is part of the Master in Industrial Mechanics in the second term of the second course. The course comprises 4 ECTS in 14 teaching weeks, so students have an hour and a half of class a week.

In this subject, students learn diagnosis of mechanical systems using vibration analysis. In this way, they learn advanced techniques for signal processing and mechanical systems identification based on data analysis. In general, the training of the students is uneven, and most of them students have never studied mechanical vibrations. Thus, sometimes it is difficult for them to consolidate the knowledge of this advanced concepts in 21 h of class.

The evolution of the number of students from 2017 to 2022 is shown in Fig. 24.1.

The weekly classes alternate theory and exercises sessions each week. The exercises constitute half of the teaching hours because they are essential to understand the advanced theoretical concepts explained. The exercise part is focused on examples extracted from [1]. The exercises are programmed in Matlalb® with friendly interfaces, showing real and visual examples to strengthen the knowledge acquired in the theory sessions.

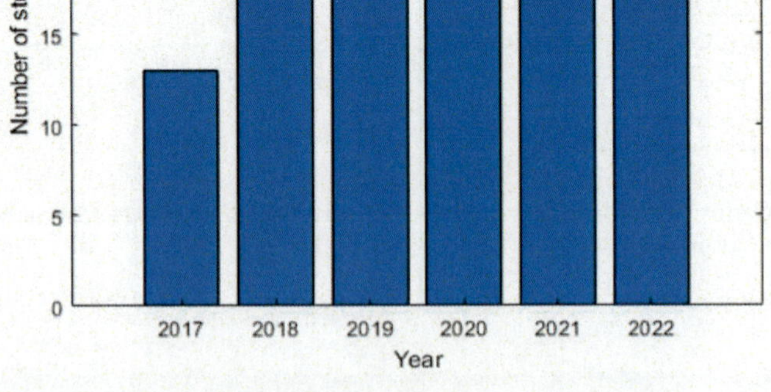

Fig. 24.1 Evolution of the number of enrolled students from 2017 to 2022

At the year 2017, it was observed during the course that most students did not participate when a question was asked by the professor after every exercise. Besides, at the end of the course the average final exam mark was low. That course, the students did a group work, which proposal was presented in [2], where they had to take vibration measurements from a fault simulation machine and perform its diagnostics based on the processing of the signals taken. It was concluded that the group work done was useful to learn how to measure, but did not help improve the understanding of the theoretical contents explained.

For this reason, and due also to the increasing in the number of students (from 13 to 27) in 2018, another type of work was proposed. The idea of the new work was to substitute the group work by an individual work, giving the students vibration signals avoiding the measurement work on their part. This way, the main work would focus in the signal processing tasks and in the diagnosis. Case Western Reserve University Database [3] signals were given to the students to perform bearing diagnostics. Using the new proposal, the final exam mark average improved in the successive years. However, it was detected that students did not follow the course up to date.

In 2020, the purely summative evaluation strategy was substituted by a new formative evaluation strategy, based on the introduction of online questionnaires for the students after each exercise or example. The value of the questionnaires' marks is 10% of the final grade.

Both strategies have shown to be positive for improving the knowledge of the students.

24.2 Methodology

To improve the understanding and the motivation of the students with the course, two different types of actions were taken:

(a) 2018: introduction of an individual work.
(b) 2020: introduction of online questionnaires.

(a) Introduction of an individual work

The substitution of a group work by an individual work was performed to avoid the consequence that only few students of the group really worked and understood the principles used.

Besides, the new individual work removed the experimental part that was required in the previous work, presented in [2], that was useful to learn how to measure, but did not prove to be as effective in learning other types of theoretical principles.

The objective of the work is for students to perform bearing diagnostics using freely available vibration signals provided by CASE WESTERN RESERVE UNIVERSITY in [3], which are widely used in research to evaluate new techniques and parameters for machine diagnostics. In this way, students are encouraged to explore the state of the art and to go beyond the techniques explained and proposed in the course. The design of the work is carried out as a script to help the student to process signals, and to extract conclusions about the optimal conditions for the diagnosis purpose.

Approximately halfway through the term, once the basic knowledge has been imparted to be able to understand and approach it, the script is shared with them during a collective tutorial. From then on, all the contents of the course are presented with the example of the application of the work, proposing reflections and debates on certain issues, as a form of interactive formative evaluation that provides valuable information to both students and teacher. In this process they are encouraged to go beyond the knowledge imparted in the course and to delve into the state of the art.

At the end of the term, students submit an individual report in which they explain the knowledge applied, reflections and conclusions obtained. This report is also finally evaluated with a numerical grade.

(b) Introduction of online questionnaires

The proposed questionnaires ask theoretical questions to verify that the students follow the class and that they understand it. It consists of a series of multiple-choice questions that help to strengthen the concepts studied in the theory classes and thus, consolidate them.

Until year 2019, these questionnaires were answered orally and jointly with all the students, so only some answered the questions, not making it possible to see who really understood it and who did not (see Fig. 24.2). The final grade of the course was not very high.

To improve both the final grade and the interest and motivation of the students, it was decided to apply these synchronous tests online to improve the participation in class, using Moodle platform. In this way, everyone must answer the questionnaires (since it counts for the final grade of the course) and force them to be attentive. It was immediately observed that the students paid much more attention, participated more and were more motivated. This allows both the student and the teacher to know in real time who has answered correctly and who has not, so it is possible to emphasize more in those concepts that people have failed the most and repeat the main concepts when necessary. With the implementation of this methodology, it has been possible to improve in numerous aspects, which are all advantages as shown in Fig. 24.3.

These advantages are:

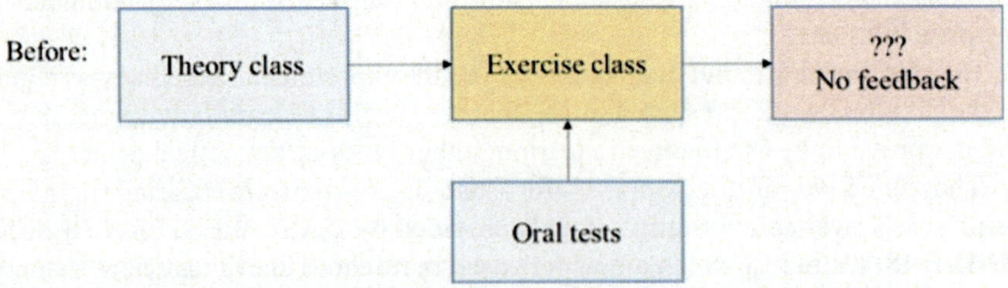

Fig. 24.2 Methodology before introducing the novelty of tests

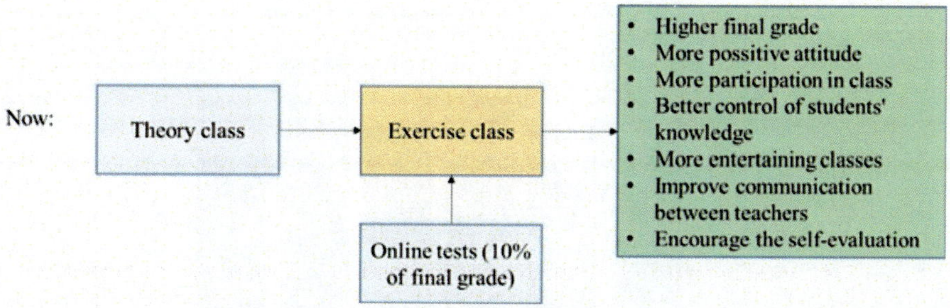

Fig. 24.3 Methodology after introducing the novelty of tests

- Higher final grade. Students demonstrated more knowledge in the final exam, obtaining higher final grade.
- More positive attitude. Students pay more attention in class and participate more.
- More participation in class. Students asked more doubts and answered the questions that the teacher asked.
- Better control of student's knowledge. Knowing in real time how many people answered the test right or wrong allows the teacher to delve deeper into the concepts that have not been clear.
- More entertaining classes. Students are attentive and alert to answer the questionnaires well and the participation of all makes it more enjoyable.
- To improve communication between teachers: digital questionnaires are accessible for all the teachers, thus all of them can follow the learning process and draw conclusions at each topic.
- To encourage the self-evaluation on the part of the student: the student is able to control the learning process and to detect misunderstandings, and gaps in knowledge.

This formative evaluation is of an interactive type, since it occurs in an integrated way with the learning process. In this way, the student regulates the learning, and on the other hand, the teacher can detect learning problems and replan strategies.

24.3 Results

To see if the actions taken have been useful for the students, the average grade of the final exam has been collected for the last years. The final mark is not considered because there have been changes in the weighting of the exam and work during the years. From 2018 to 2021, the new individual work was done. In 2022 the group work proposed in [2] was done again. Regarding the questionnaires, they have been used from 2020 to 2022.

Figure 24.4 shows the evolution of the final exam grade average (out of 10) at the first call from 2017 to 2022. It can be observed that the substitution of the group

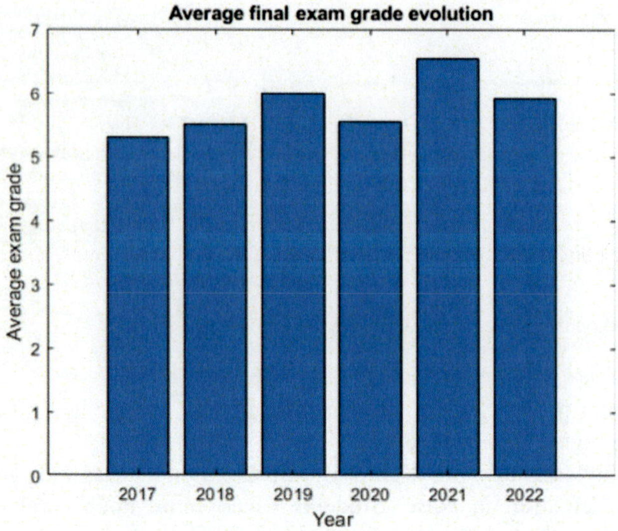

Fig. 24.4 Evolution of the final exam mark average (out of 10) with the actions taken

work by the individual work from 2018 to 2021 helped understanding the theoretical concepts and improving the final exam marks, that decreased again in 2022 when the group work is introduced.

Another consequence that can be obtained from the actions taken is that the number of students that attend the final exam at the first call have improved considerably with the improvement of their knowledge and understanding, reaching values of 100% from the introduction of the questionnaires in 2020, as can be seen in Fig. 24.5.

24.4 Conclusions

The implementation of the actions has demonstrated to be very useful for the development of the course.

Students are more participative and the knowledge acquired is better. The introduction of the tests has allowed their self-evaluation, and encouraged them to keep up to date. Since the implementation of the tests in 2020, 100% of the students attend the exam at the first call. The performance of an individual or a group work also affects the final exam mark, seeing that when students do an individual work, the understanding of the concepts is better, avoiding that only few students of the group really work.

In general, many improvements have been detected in the students' attitude and training.

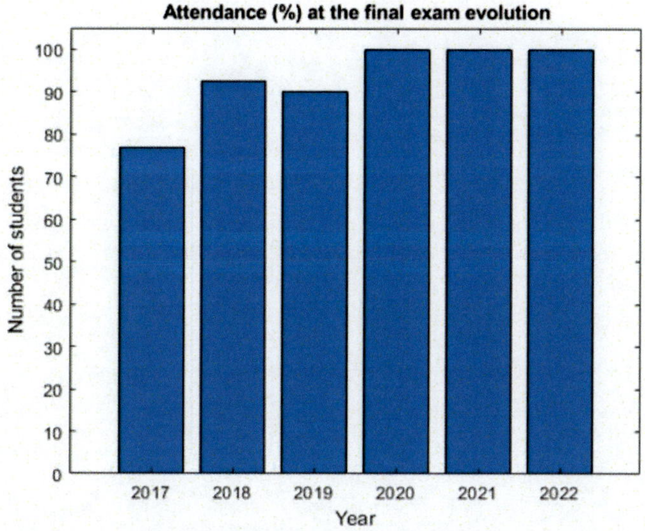

Fig. 24.5 Evolution of percentage of attendance at the first call exam

References

1. Braun, S.: Discover Signal Processing: An Interactive Guide for Engineers. Wiley (2008)
2. Zamorano, M., Gómez, M.J., Castejón, C., García-Prada, J.C.: Learning machine diagnostics through laboratory experiments. In: New Trends in Educational Activity in the Field of Mechanism and Machine Theory, pp. 57–63. Springer (2019)
3. Bearing data center. Case western reserve university database. https://engineering.case.edu/bearingdatacenter. Accessed 29 Sept 2022

Index

J. C. García Prada et al. (eds.), *Trends in Educational Activity in the Field of Mechanism and Machine Theory (2018–2022)*, Mechanisms and Machine Science 128,
https://doi.org/10.1007/978-3-031-25730-8

M

N

O

P

R

S

T

V